Every night the twin domes of the Keck Telescope examine the depths of space from the summit of the extinct volcano Mauna Kea, the highest point in the Hawaiian Islands. Their principal mirrors are a record 33 feet (10 meters) in diameter, which allows them to capture light from extremely faint bodies. These mirrors were not made from one enormous block of polished glass, but were assembled from a mosaic of small polygonal mirrors that are much easier to manipulate than a single piece

COSMOS

Sylvia ARDITI - Marc LACHIÈZE-REY

FIREFLY BOOKS

FIREFLY BOOKS

Published by Firefly Books Ltd. 2004

English language edition copyright © 2004 Firefly Books

All rights reserved. No part of this publication may be reproduced, stored in a retrieval system, or transmitted in any form or by any means, electronic, mechanical, photocopying, recording or otherwise, without the prior written permission of the Publisher.

Publisher Cataloging-in-Publication Data (U.S.)

Arditi, Sylvia.
    Cosmos / Sylvia Arditi ; Marc Lachièze-Rey.
Originally published: Paris: Marval Vilo, 2003.
[176] p. : col. photos. ; cm.
Includes bibliographical references and index.
ISBN 1-55297-932-6
1. Cosmology. 2. Astronomy. I. Lachièze-Rey, Marc. II. Title.
523.1 dc22 QB982.A73 2004

National Library of Canada Cataloguing in Publication

Arditi, Sylvia
    Cosmos / Sylvia Arditi, Marc Lachièze-Rey.
Includes index.
ISBN 1-55297-932-6
    1. Astronomy--Popular works. I. Lachièze-Rey, Marc II. Title.
QB44.3.A73 2004   520   C2004-901916-3

Published in the United States in 2004 by
Firefly Books (U.S.) Inc.
P.O. Box 1338, Ellicott Station
Buffalo, New York 14205

Published in Canada in 2004 by
Firefly Books Ltd.
66 Leek Crescent
Richmond Hill, Ontario L4B 1H1

Printed in Spain

First published by Vilo Editions in French, 2003

Design: PATRICK REMY STUDIO

Editing: BOWER

Illustrations: CIEL ET ESPACE

Graphic Design and Photoengraving: MAESTRO

Translated by: Global Translations and Interpreters Services Inc.
www.globaltranslations.com

TABLE OF CONTENTS

SPACE-TIME IN EXPANSION AND DEVELOPMENT

Four centuries BC, Plato emphasized the harmony of the world. He called it cosmos from the Greek term for "harmony." He had no idea of the variety of splendors that would be unveiled more than 2,000 years later by the telescopes and other instruments that examine the universe today—the subject of this book.

THE FIRST EPOCH

Plato and his disciple Aristotle founded the first epoch of our civilization's cosmology. The world that they described is very different from ours. Its entire extent corresponds to what we today describe as the Solar System, the subject of Part One of these photographs.

Above all else, that world was hierarchical. It had a center that coincided with the Earth's. It was enclosed by a spherical boundary, called the sphere of fixed objects. Between the two extremes, nested celestial spheres carried the Moon, the Sun and the planets.

The first orbit, the Moon's, established a fundamental distinction within this system. The sublunary world—essentially our planet and its atmosphere—constituted the domain of the perishable, in which everything changed, decayed and ended by dying. What existed here corresponded to combinations of four elements (earth, water, air and fire), combinations that formed and broke down without end.

In contrast, in the world beyond the Moon nothing ever changed—everything remained the same. There were no traces of the four elements, for heaven was filled with an ethereal substance called quintessence (the fifth essence), reflecting the perfection and harmony that was lacking on our planet. Whereas on Earth movements were all vertical (toward the center), in the Platonic cosmos the sphere and the circle reigned, geometrical figures reputed to be perfect and harmonious. People studying geometry and astronomy described these celestial movements using extremely complicated combinations of circles and spheres, notably in the second century BC in Claudius Ptolemy's famous epicycles. This model reigned for 15 centuries—a longevity unequaled in the history of science—until Kepler introduced ellipses.

STELLA NOVA

After much delay and hesitation, Christian antiquity came to accommodate the cosmology of Plato and Aristotle, as interpreted in conformity with the Scriptures. It was truly a unique and enclosed world that God had created for Man, with our planet positioned at the center. But in the 16th and 17th centuries a series of challenges destroyed this world view. They prepared the way for the second epoch of our cosmology, which would be marked by the publication of Newton's *Principia*.

Copernicus's displacement of the center of the universe to the Sun (1547) was considered unacceptable by the Church. Even more unacceptable was the existence, affirmed a little later by Giordano Bruno, of many other inhabited worlds. On February 17, 1600, after having had his tongue cut out for "the horrible words that he had uttered," Bruno was burned in Rome as a heretic. His declarations nonetheless greatly served to spread the ideas of the Greek atomists from well before Plato and Aristotle. Bruno was an ardent defender of the idea of infinity, which he linked to the divine power by "the argument of plenty," namely, that it would limit the concept of an all-powerful God to think that He would restrict creation to a finite world. The dogmatists were not yet swayed, but the astronomy of Ptolemy, built on the Aristotelian system, was showing more and more weaknesses.

The Danish astronomer Tycho Brahe built very precise instruments without having either lenses or telescopes. By observing the heavens in great detail, he demonstrated the inadequacies of the Alfonsine Tables, which predicted the positions of heavenly bodies by using the ancient Ptolemaic cosmological system. These observations allowed Johannes Kepler to replace them with the Rudolfine Tables (1627), named in homage to Emperor Rudolf II. Prudently, Tycho Brahe created a system halfway between that of Aristotle and the heliocentrism of Copernicus, a system that could satisfy the Church—the Moon and Sun orbited the Earth, but the planets orbited the Sun.

In 1572 Tycho Brahe observed a "new star." Was it possible for the heavens to change?

Johannes Kepler had been born the previous year. He also observed a new star, in 1604, which he wrote about in his *De Stella Nova*. Aristotle's unchanging heavens could not support being confronted with these new objects. Today we know that these new stars were supernovas, violent astronomical events that mark the end of life for stars; the huge increase in their luminosity makes them visible on Earth (see illustrations on pages 120 to 139).

Stars with Tails

In 1680 and 1682, two successive comets appeared. Tycho Brahe and Johannes Kepler speculated about the passage of these heavenly travelers (see illustration on page 44). A meticulous astronomer, Tycho Brahe undertook to measure their distances, and succeeded in establishing that they were moving beyond the lunar orbit. Since they were not in the sublunary environment (our atmosphere), but much farther off, the world beyond the Moon could no longer be considered as unchanging.

In his *Principia* (1687), Isaac Newton set forth the laws of dynamics and gravitation that would explain the movements of the comets. From then on it would be possible to predict the return of these symbols of the unforeseeable, whose unexpected appearances had caused terror among the people.

Finally, still working from the results of Tycho Brahe's observations, Kepler showed in 1605 that the orbit of the planet Mars is elliptical. And so he put an end to the era of the sacred pre-eminence of the circle in the realm of the planets.

A little later, in Venice, Galileo Galilei improved on the telescope invented a short time earlier, and turned it toward the heavens. What he discovered was astounding. In 1610 he wrote about it in his *Siderius Nuncius* [The Starry Messenger]— mountains on the Moon as on Earth; phases of the planet Venus, analogous to those of the Moon; satellites around Jupiter; spots on the Sun; and countless new stars unseen by the naked eye. That was already a lot. At the same time, Galileo started to understand the mechanisms of kinematics and dynamics, the principles of motion for which Newton would write the laws. Like his successors, he soon concluded that the separation between the two worlds—below and beyond the Moon—was over.

With one challenge after another, the Aristotelian concepts collapsed. A new physics, a new astrophysics and a new cosmology took their place. Newton's *Principia* is often considered their official birth certificate.

On Earth as It Is in Heaven

Everything was turned upside down. It was the end of the Aristotelian world. In many respects, such as ideas about the composition of matter and of infinite space and many worlds, the concepts moved closer to the views of the ancient atomists.

Astronomers came to understand that distances to the stars varied, and exceeded the confines of the Solar System. The world was no longer enclosed. It became the Universe, unique and unified, and the composition of matter, motion and the laws of physics were the same everywhere, below and above. In the 19th century, the invention of spectroscopy confirmed the nature of matter, both on our planet and in the stars, by showing the common properties of the light they emit.

The pre-eminence of vertical motion on Earth was no longer an expression of a fundamental fact (as opposed to the heavenly reign of circles and spheres), but entirely contingent on the unified kinematics and dynamics whose laws Newton had set down. The laws were truly universal. Those of gravitation accounted for both falling bodies on Earth (in Newton's case, apples) and orbiting heavenly bodies. Newton's laws also explained the laws discovered by Kepler. Even comets obey the universal rules.

On Christmas Day 1758, a German farmer reported the approach of a comet. The calculations of Newtonian physics had predicted that Halley's Comet, already observed in 1531, 1607 and 1682, would return at that time. Verification of this prediction overcame the last resistance toward Newtonian theory and assured its immense success.

At the beginning of the 19th century, astronomical calculations did not agree with the observed orbit of the planet Uranus. Using Newton's laws, the Parisian astronomer Urbain-Jean-Joseph Le Verrier proposed the existence of an disruptive unknown planet, whose position he provided. In September 1846, the astronomer Johann Gottfried Galle, at the Berlin Observatory, discovered a luminous point in the direction calculated. The discovery of Neptune was the crowning moment of Newtonian theory, a theory that works so well it stands as a model to inspire other disciplines.

INFINITE SPACE: THE SECOND EPOCH

Newton immeasurably extended the limits of the world by incorporating the Euclidean geometry of space into the framework of physics. Infinite, geometric and homogeneous (that is, the same throughout), the new world had nothing in common with the ancient. The old ideas of the atomists were rehabilitated—after Newton, no physicist and no astronomer seriously doubted the multiplicity of worlds. Almost everyone speculated that the other planets harbored creatures more or less like us.

But after centuries and centuries of horror of the infinite, the idea was not easy to digest. Further, it went against the paradox of the dark night sky—in an infinite space uniformly filled with stars, should not the night sky remain uniformly brilliant? Astronomers also undertook to determine the limits of the material world in this infinite space. How is this world organized? How does it occupy space? That required estimating the distances to the stars, an extremely difficult task that progressed only little by little. It would lead finally to a new shock—the dawn of the third cosmological period at the beginning of the 20th century.

A PERFECT BUT INCONCEIVABLE WORLD

Despite the immense success of its predictions, Newtonian physics still left a few unsatisfactory holes. All motions, terrestrial or celestial, had been perfectly predicted and explained, but this avalanche of successes masked conceptual difficulties.

Newton invoked gravitation as a force, an interaction at a distance. But how could two bodies influence (that is, attract) each other without touching? Newton's contemporaries had difficulty accepting action at a distance. Newton himself, while aware of the difficulty, remained powerless to address it. His theory worked—that was already very satisfying. Some tried to invoke "ether," a sort of elastic, invisible medium filling the universe, whose role would be to transmit gravitation. It was a little like space, with additional physical properties. But ether remained paradoxical and resisted all theoretical explanation. Critics of that concept went back to the concept of space. But how could one base a theory on this *space*, which no one could see or touch, a "pointless metaphysical entity," according to Ernst Mach, one of the principal commentators?

To this difficulty was added another. Gravitation, as proposed by Newton, showed a remarkable property, already understood by Galileo, that was expressed in the form of what is called the *equivalence principle*. That is, subjected to the same gravitational interaction, all bodies, whatever their properties, experience the same acceleration. Newton's theory takes this property into account. But it allows the mystery to continue, because it does not explain why it should be that way.

These questions fed the reflections of philosophers and physicists for three centuries, during which the successes of Newtonian physics accumulated. In the 19th century they would intermingle with questions about the propagation of light. At the beginning of the following century, young Albert Einstein provided the solution. Whereas Newtonian physics and electromagnetism covered almost all observed phenomena, Einstein's purely conceptual reflections led him to propose two new theories—special and general relativity. Developed in response to earlier conceptual weaknesses, his theories turned out to have a concrete applicability superior to that of Newtonian physics.

As we know, relativity united the two notions of space and time into a single concept—space-time. It also used new, "non-Euclidean" geometries (spherical, for example) discovered by mathematicians in the 19th century. Each point in space-time is characterized by a geometric quantity called the *curvature tensor*, which expresses its deformation. On one hand, this curvature defines the "shape" of space-time at the point under consideration. On the other hand, it expresses the gravitation at that point, considered not as a force, but as an expression of the deformation of space-time. This new theory, still more geometric than Newton's, resolved all the objections directed at its predecessor.

A RELATIVISTIC COSMOS: THE THIRD EPOCH

General relativity revolutionized physics in a way that undoubtedly has not yet been fully assimilated by the physics community of today. At the same time, it paved the way to the third epoch of cosmology.

At the end of the 18th century, the German philosopher Immanuel Kant put forth the hypothesis of "island universes," that matter may be distributed in the cosmos like islands in an archipelago emerging from the dark sea. What we see around us might be only one of many island universes. To develop the concept more clearly, observers of the heavens explored the structure of the universe by estimating the distances to the various stars they observed. Their results became more precise, and since the 19th century astronomers have been able to declare that the stars we observe are distributed in a flattened disk of astronomical dimensions indeed—about 50,000 light-years in diameter, or a million times the diameter of the Solar System. The most distant stars in this disk are not visible to the naked eye, but their light builds up to form a luminous band across the sky—the Milky Way. In the 19th century this gathering of stars, our galaxy, represented the entire extent of the material world in Newton's infinite space.

But lenses and telescopes revealed other objects in the sky besides stars, most notably nebulas. The astronomer Charles Messier prepared the first catalog of nebulas in 1781. The invention of photography made it possible to observe them in more detail, and some nebulas attracted people's interest because of their characteristic spiral form. At the end of the 19th century the American astronomer Vesto Slipher used spectroscopic methods to measure their speeds, which turned out to be extraordinarily high—some hundreds of thousands of miles per second. Slipher deduced that these objects were moving so quickly that they must soon leave our galaxy, and therefore suggested that they were located outside of it. Could these be Kant's island universes? Accepting that theory would lead to acknowledging that the universe far exceeded our own galaxy in extent. The arguments on both sides grew fierce, and the situation was as confused as the observational results were contradictory. This great debate came to an end in 1924, when Edwin Hubble, another American astronomer, showed that Andromeda, the most brilliant of the nebulas, was definitely located outside the limits of our galaxy. Not only is it not a cloud in our own galaxy, it is another, totally separate spiral galaxy (see illustrations on pages 140 to 158).

A GAS OF GALAXIES

This dramatic turn of events pushed the limits of the world out even farther; galaxies, including our own, become like atoms in a cosmic fluid of immense vastness. Astronomers today have identified millions of galaxies, and they estimate that very likely there are billions. Those discoveries were made within just a few years of the theory of general relativity. Its creator, Einstein, like some of his contemporaries, rapidly came to understand that it would revolutionize the approach to cosmology. A new vision of the world was emerging to take into account these new results.

The biggest shock was yet to come. In 1924 the source of the velocities measured by Slipher remained an enigma. Further, he had measured even more nebulas; their velocities were even higher, and they were always receding, never approaching. The scientific community unanimously recognized a phenomenon on a cosmic scale, but got caught up in trying to explain it. The Belgian physicist Georges Lemaître was the first to recognize what is today called the expansion of the universe, and showed that it constituted a natural solution to the equations of general relativity. (In 1922 the Soviet scientist Aleksandr Friedmann also discovered this solution, but he did not make the connection with observations of the velocities of galaxies.) This new upheaval triggered nearly as many negative reactions as the Copernican revolution. Several decades were needed before it was finally accepted without hesitation. It would take a little longer for the model of the Big Bang, also introduced by Lemaître (at first as "hypothesis of the primeval atom"), which described an evolving universe. It is difficult to attack the old myths—here, that of a static universe. Even though Lemaître was not burned as a heretic, he did suffer indifference and mockery. One of his principal adversaries derided him as the "Big-Bang man"; the term was particularly poorly chosen, but it stuck.

Our universe today is an expanding and evolving space-time. Models of the Big Bang describe its history over 15 billion years—the expansion and cooling of its contents and the long development of its structure. First, elementary particles appeared, and then atomic nuclei and the atoms themselves. Then came molecules and astronomical structures—stars, planets, galaxies and more. Matter is confined within structures such as planets in the Solar System, and probably in innumerable other stellar systems; stars in galaxies; galaxies in clusters. The latter themselves form gigantic cosmic structures whose expanse is measured in millions of light-years; these are the largest systems that we know of in the cosmos.

At every scale, telescopes reveal to us a richness and variety of structure that our predecessors could not even suspect—a pleasure for admirers of the sky, and a challenge for astrophysicists.

— *Marc Lachièze-Rey*

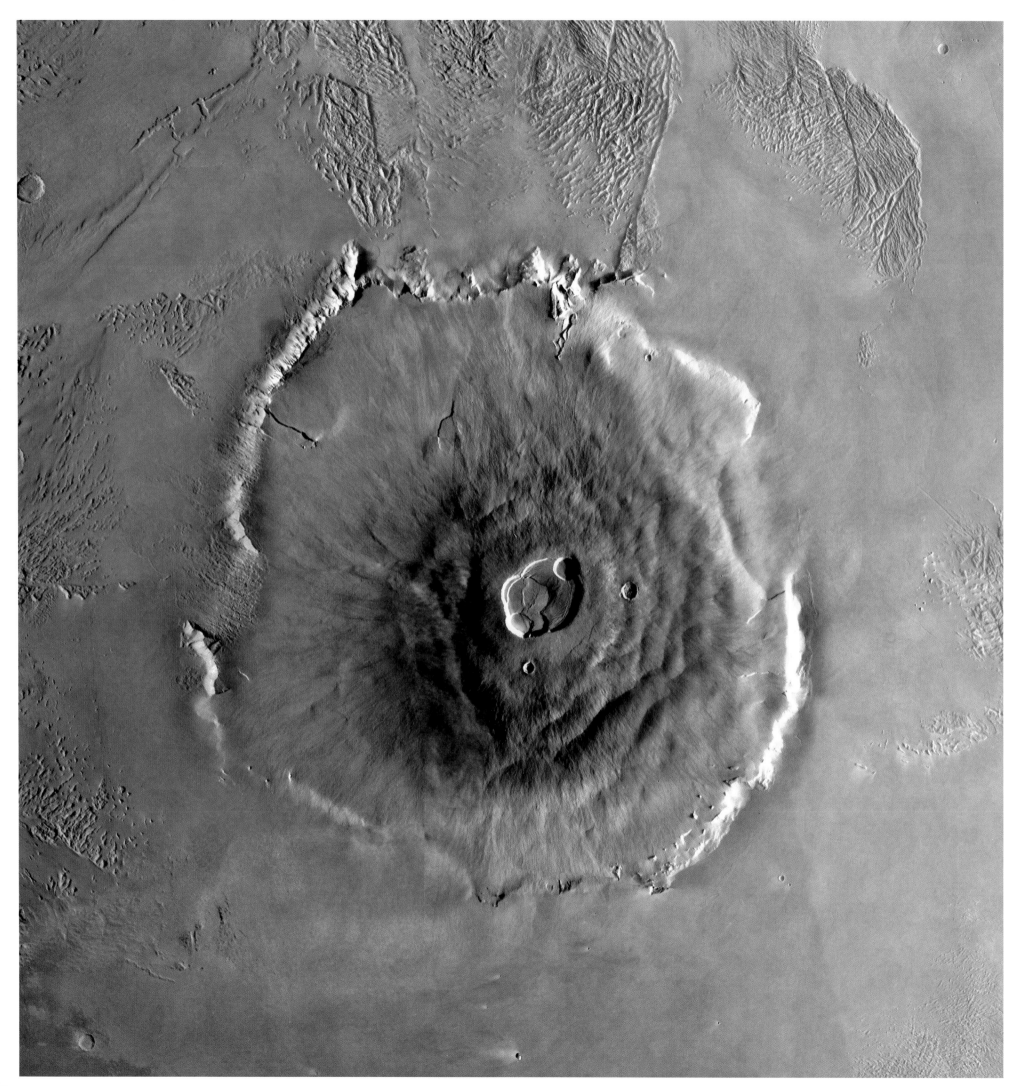

THE COSMOS NEARBY: OUR SOLAR SYSTEM

Our star carries along with it on its path a procession of nine planets and 61 moons, with tens of thousands of rocks and chunks of ice and an incalculable number of grains of dust, all born 4.5 billion years ago in a vast cloud of gas and dust in the Milky Way. This cloud collapsed while rotating around itself. At its center, gaseous matter heated to a Dantesque temperature formed an incandescent sphere—the Sun.

At the edges, minuscule solid bodies fused together to form planets, moons and wandering bodies, the comets and asteroids. These meteors, traveling in all directions, crashed down in a rain of fire and ice on still-warm planets and moons, digging numerous craters.

Some of these worlds have scarcely changed since their formation. Others have undergone tumult, noise and fury. All of them differ—icy minerals here, suffocating furnace there; calm and monotonous or raging cataclysm; light breezes or raging hurricanes…

Olympus Mons on Mars

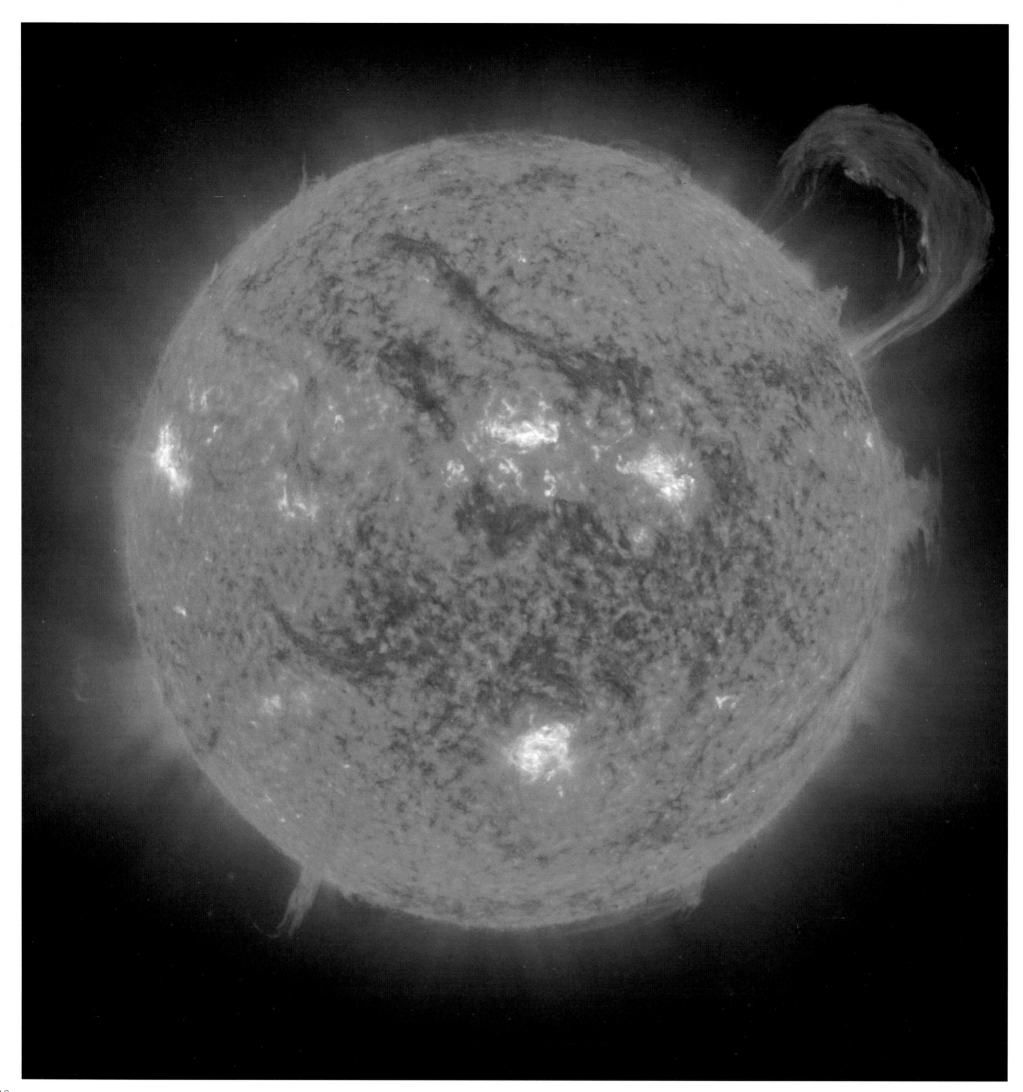

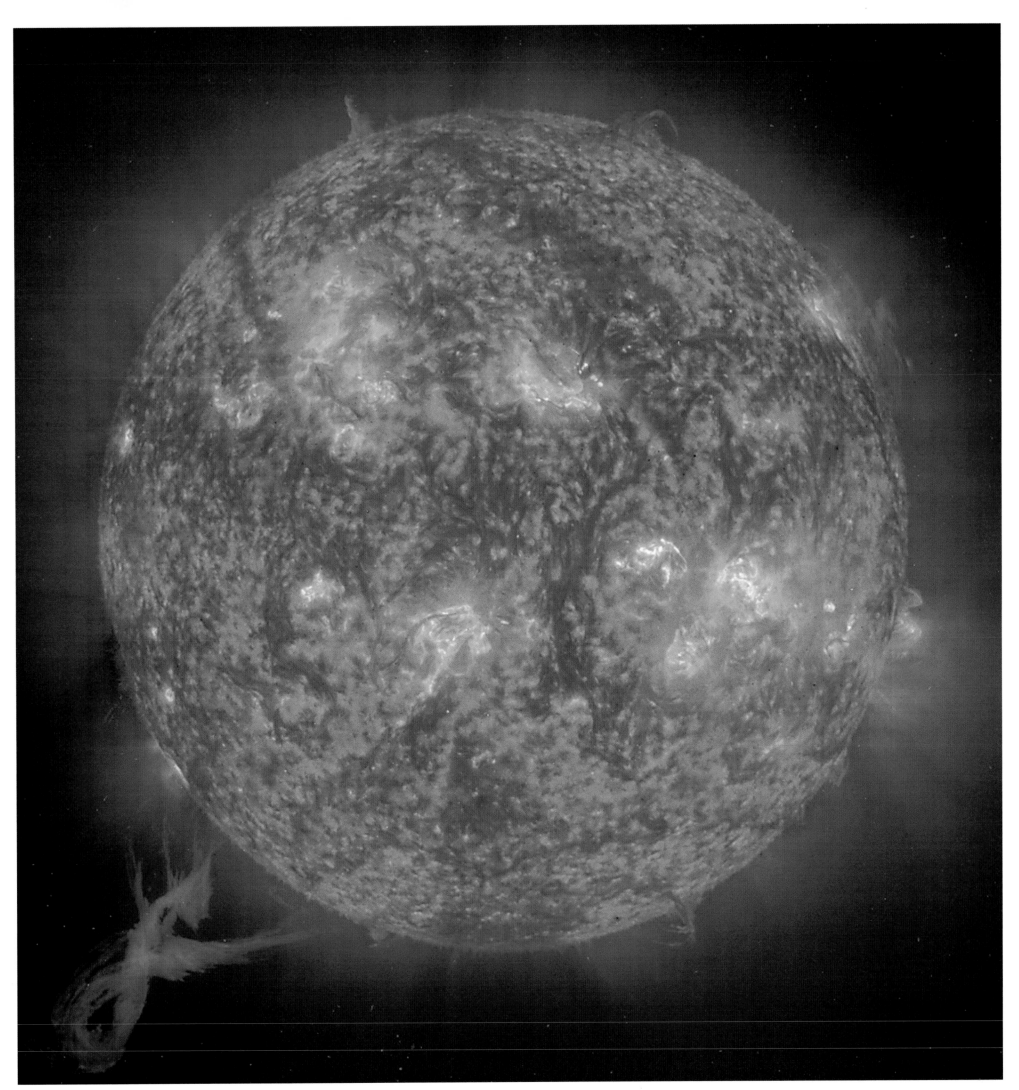

OLYMPUS MONS ON MARS
(VIKING)

No one would have wanted to live at the foot of Olympus Mons, at least not while it was active. It is the largest volcano in the entire Solar System, deforming the surface of the fourth planet, Mars. It is a cone with a gentle slope, 435 miles (700 km) in diameter and with a summit that reaches an altitude of over 17 miles (27 km)—three times higher than Earth's greatest volcanoes, the Hawaiian Islands. At the summit, the caldera, or crater, spreads wide a gaping mouth that is almost two miles (3 km) deep. The formation of this volcano, caused by the piling on of successive layers of lava, was made possible because, among other reasons, the crust of Mars does not move around as Earth's does. Today the volcano is inactive, as are the other Martian volcanoes. About 800 million years ago the radioactive compounds trapped in the core of Mars were spent, no longer reheating its magma. The interior of the Red Planet then solidified, and its volcanoes were extinguished for eternity.

Mars is no longer anything more than a desert frozen at –40°F (–40°C), its sterile soil covered with dust and loose stones colored red by rust (iron oxides). Despite all this, astrophysicists assume that this planet long ago had a more temperate climate and that liquid water probably flowed on its surface. How did it meet such a dismal fate, so different from Earth's? The first explanation that comes to mind is that Mars is too far from the Sun and does not receive enough heat. But it turns out that solar heating does not entirely explain the differences. Other parameters play an essential role, starting with the composition of the atmosphere. If Earth were without an atmosphere, its average surface temperature would be about zero Fahrenheit (–18°C), not 60°F (15°C). The atmosphere acts like a cocoon, retaining the heat coming from the Sun instead of letting it escape into space. The intensity of this greenhouse effect depends on, among other things, the quantity of CO_2 (carbon dioxide) in the atmosphere.

The atmospheres of Earth, Venus and Mars originated a little after their formation 4.5 billion years ago, when volcanoes ejected the gases contained deep inside them. Water vapor was among these ejected gases, and it condensed and fell back to the ground in the form of torrential rains, little by little forming oceans, and perhaps rivers. During this time CO_2 was also spewed out by the volcanoes, creating a greenhouse effect that prevented the water from freezing. Biologists and planetologists do not reject the possibility that Mars might have given rise to living organisms, as water may have been flowing on the surface for long periods. But Mars is a small planet, with a diameter half that of Earth's and a lower density. Consequently, its gravitational attraction—one-third that of Earth's—is not sufficient to retain atmospheric gases for long. These gases, as well as the water, therefore ended up evaporating into space, although some water seems to have seeped into the ground before freezing in place. Although rich in CO_2, the Martian atmosphere is actually extremely thin; its pressure is a hundred times less than that on the surface of Earth.

In the years to come, more and more sophisticated robots are going to land on this planet to dig in the soil, searching for traces of water and past life. Perhaps they will reveal to us the secret of the Red Planet. Or will it perhaps be necessary to wait for the definitive answer until, at last, humans dig on the surface?

THE SUN
(SOHO)

The impassive face that the Sun, our star, offers us is only an appearance. It is a ball of gas—hydrogen, helium and traces of heavy elements—brought to incandescence and ionized into plasma. This sphere emits not only light and heat. Every moment it is hurling into space, out to Earth and the other planets, shreds of gaseous matter with the power of several billion thermonuclear bombs. This continuous torrent of charged particles—protons, electrons and heavy ions—has gradually reduced the rocks on the surface of Mercury and the Moon to powder. Our planet is protected by a magnetic shield that moves the charged particles toward the poles or shunts them aside. Periodically (roughly every 11 years) this solar wind is unleashed and blows in violent gusts. The consequences can be spectacular; for example, the solar wind tripped the entire electrical grid of the province of Quebec on March 13, 1989, disabling it for six hours.

The phenomena at the source of these disturbances can now be observed. These two images, captured by the American and European spacecraft SOHO (Solar and Heliospheric Observatory) on September 14, 2000, and January 18, 2001, bear witness to the violent ejection of matter in the chromosphere, the gaseous layer 1,250 miles (2,000 km) thick that envelops the visible surface of the Sun. Two gigantic plasma protuberances, which could each contain several tens of Earths, are clearly visible at the top right of the first image and at the lower left in the second. They shoot up to several hundreds of thousands of miles above the surface, equivalent to the distance between Earth and the Moon. Some of these arcs of matter remain suspended for several days, even months, on end; others, more unstable, explode and release a torrent of charged gas in a solar eruption. Flares in the direction of Earth can let loose electromagnetic storms with the consequences described above.

These images, captured in extreme ultraviolet (here, waves emitted by ionized iron), show among other things that the disk itself is far from homogeneous. The coolest zones (11,000°F / 6,000°C) appear in dark shades; the hottest are lighter. Within the hot zones the temperature climbs to 108,000°F (60,000°C).

The visible structures in these images are linked to solar magnetic fields. The Sun's magnetic field is ten thousand times more powerful than Earth's, but, above all, much less stable. Its polarity shifts every 22 years, and in that interval it acquires a very complex structure. Instead of emerging from the star at the two poles, the magnetic field forms multiple loops that burst out and then dive back below the surface of the star in the areas of dark sunspots (not seen here). The movements of ionized gas in the chromosphere follow these trajectories. Much research will be necessary, however, to better understand the formation of these structures and to predict, even if only hours in advance, eruptions directed toward our planet. That would allow us to protect the most vulnerable electronic equipment from magnetic storms and to prevent airplanes from flying over the polar regions during the strongest "gusts."

The prodigious energy at the source of the solar wind is also what heats the Sun and makes it shine. Scientists in the 18th century imagined that the Sun was a giant piece of coal burning itself out. They had to abandon that hypothesis when it was discovered that combustion requires oxygen, which is absent in the interplanetary environment. Further, it was calculated that a piece of coal the size of the Sun would be consumed in 30 years. The mystery was solved in the 1920s with the theoretical discovery of nuclear fusion—four hydrogen nuclei (protons) could be joined to form a lighter helium nucleus, liberating a finite quantity of energy. But it was still unclear how protons, which repel each other because they bear a positive charge (electric charges of the same sign repel), could be joined together. The answer was found during the 1930s, when it was shown that at a temperature of 15 million degrees, protons collide so violently that they can remain stuck together.

The Sun will continue to shine until it has depleted its reserves of hydrogen—in about five billion years. Then it will swell to an immense size and become a red giant star. This will engulf Earth and, once it has expelled its gaseous envelope into space, the star will be transformed into a black dwarf—a pale, insignificant stellar cadaver.

THE SUN IN NUMBERS

Mass: 330,000 Earth masses, or two billion billion billion tons
Diameter: 870,000 miles (1.4 million km), or 110 Earth diameters
Distance to Earth: 93 million miles (150 million km), or 1 astronomical unit (AU)
Temperature: 27 million degrees Fahrenheit (15 million degrees Celsius) at the center; 11,000°F (6,000°C) at the surface
Rotation: 25 days (equatorial regions); 35 days (polar regions)

PAGES 16 AND 17

CLOUD COVER AND SURFACE OF VENUS
(PIONEER VENUS AND MAGELLAN)

Venus, the Morning or Evening Star, is in reality a planet, the second closest to the Sun, and not a star at all. It owes these names to its brilliant appearance just before sunrise and just after sunset, because it is close to the Daytime Star (the Sun) and reflects its light. Venus is actually covered by a thick layer of clouds. The image on the left, transmitted by the Pioneer Venus orbiter spacecraft on February 26, 1979, shows that they completely conceal the planet's surface.

For a long time Venus made astronomers dream. What if this planet were swarming with life, a humid and lush tropical Eden cocooned under the thick cloudy layer? Unfortunately, the evidence has proved otherwise—Venus today does not provide suitable conditions to maintain any form of life, even microscopic. The planet is a true oven. It averages 860°F (460°C), hotter even than Mercury, though Mercury is closer to the Sun. Its gloom is almost complete, and the dark clouds release rains of corrosive sulfuric acid. No astronaut could survive there.

To obtain a detailed image of the surface of Venus, it was necessary to wait for the Magellan spacecraft, which from 1990 to 1994 mapped the planet using a radar technique with an accuracy of 400 feet (120 m). Venus was shown to be pierced everywhere by gaping volcanic mouths. The most imposing of these, visible as brighter spots in the photograph, have diameters of 95 to 220 miles (150 to 350 km); the basaltic lava flows are clearly visible. These volcanoes are responsible for the highly inhospitable character of Venus's atmosphere. They have never really ceased being active, and are polluting the atmosphere with their greenhouse gases (95 percent CO_2); this crushing murk weighs as much as a layer of water more than 3,000 feet (1,000 m) thick. It is possible that the planet experienced a more hospitable period 300 million years after its formation, when water vapor spewed out by the volcanoes condensed into oceans like those on Earth. But the temperature rose, the water started to boil, and every drop evaporated. It is unlikely that life would have had time to develop there.

VENUS IN NUMBERS

Mass: 0.81 Earth masses
Diameter: 0.95 Earth diameters
Distance to the Sun: 0.7 AU
Temperature: up to 900°F (480°C) on the surface
Rotation: 243 Earth days
Revolution: 225 Earth days
Number of Satellites: 0
Distinguishing Marks: intense volcanic activity

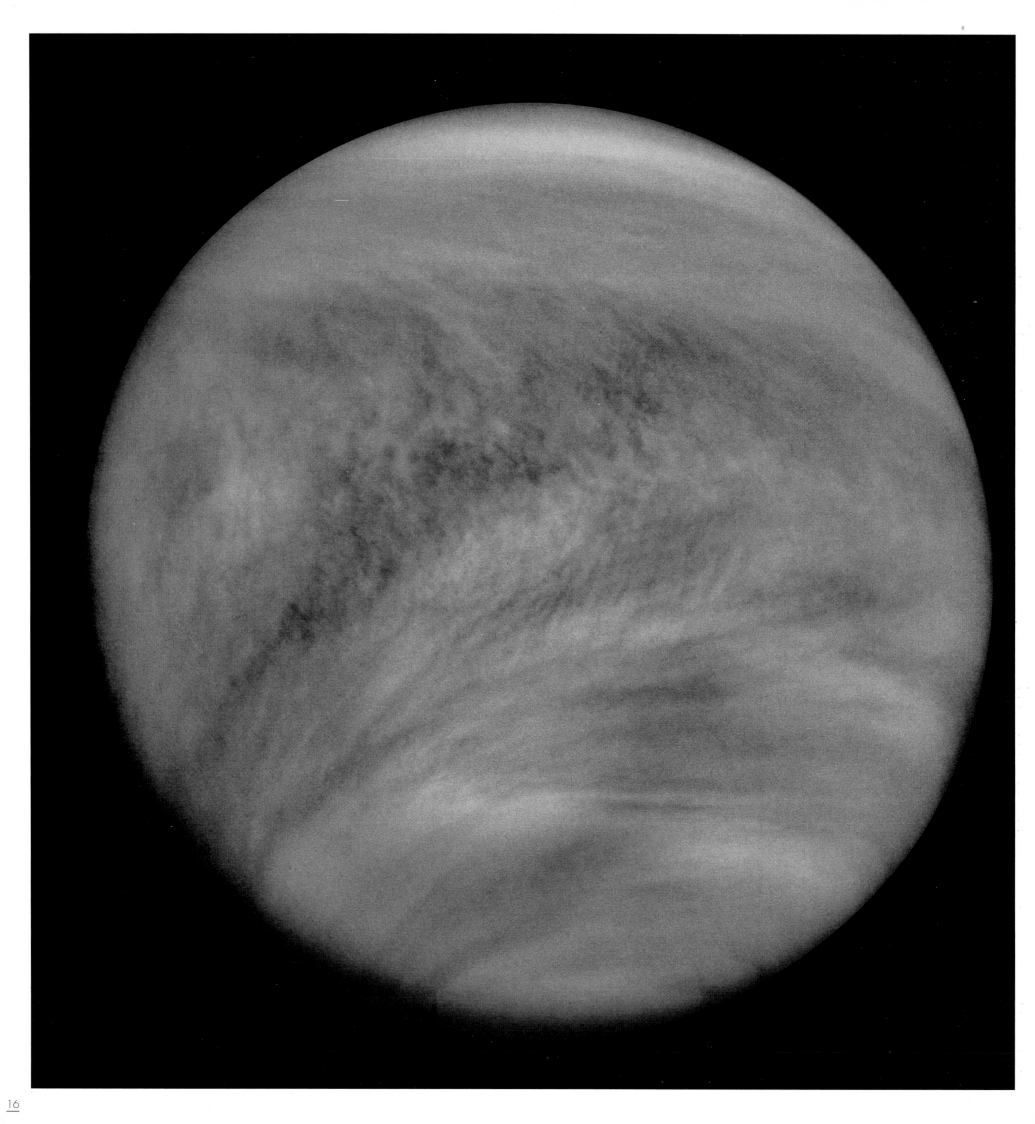

Earth's atmosphere interferes with astronomical observations, even from places with a reputation for the exceptional clarity of their sky. To escape beyond the atmosphere, a space telescope named Hubble was placed in orbit around Earth in April 1990. Its beginnings were hardly promising—the optical systems suffered from myopia. Since its repair, the telescope has been providing us with an unending flow of extraordinarily accurate images of all types of celestial objects, from planets in the Solar System to the most distant galaxies and, in between, nebulas giving birth to new stars.

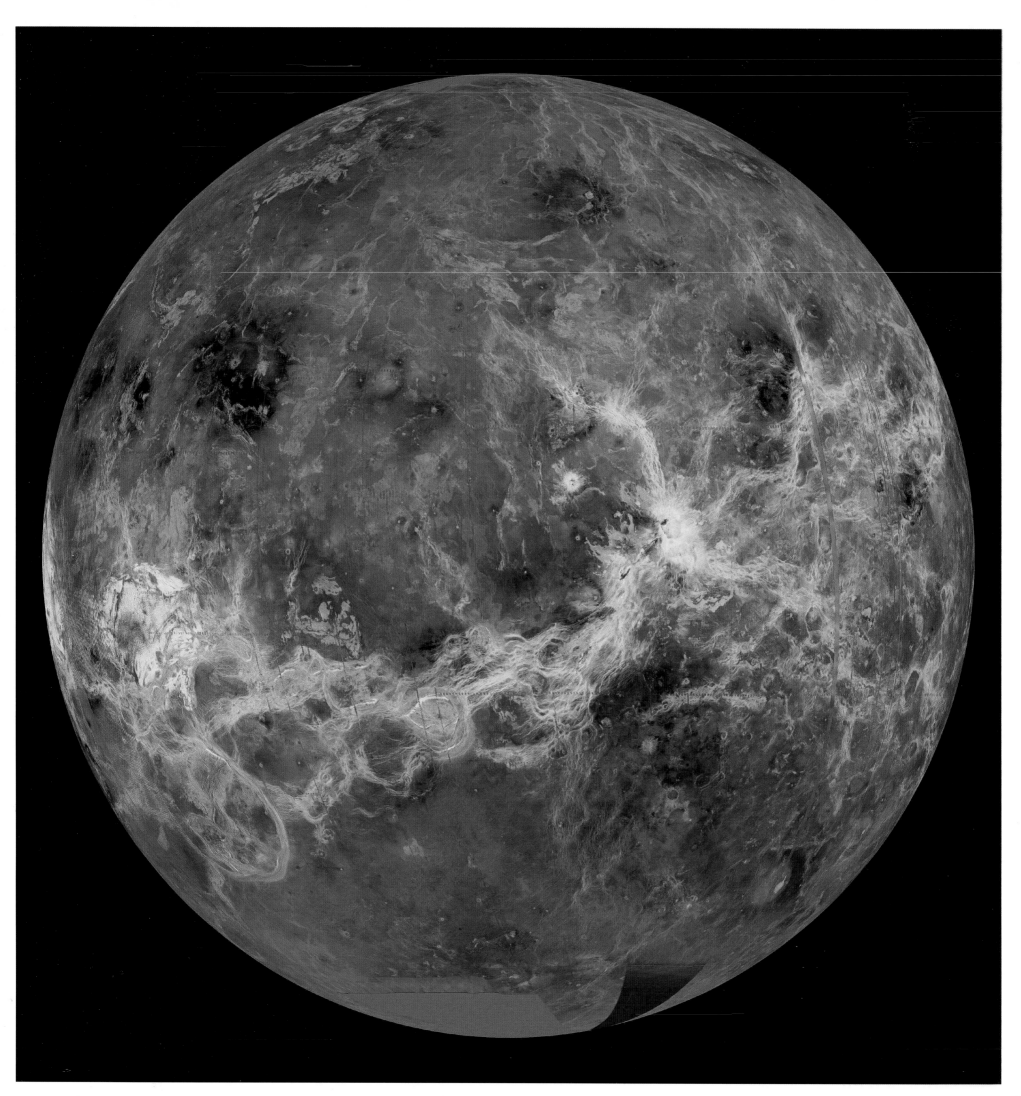

HIDDEN SIDE OF THE MOON
(CLEMENTINE)

The Moon, whose hidden face is seen here in a mosaic of images, may be the daughter of Earth. That is at least the most plausible of four hypotheses considered since the 19th century to explain its presence. For the astronomer George Darwin, son of the celebrated naturalist, the Moon may have been detached from Earth just after it was formed, when it was still liquid. Others think it may have formed independently of Earth, but nearby, or that it clumped together somewhere else in the Solar System before drawing close to our planet. According to the most recent hypothesis, proposed in 1975, a body the size of Mars collided with Earth shortly after its formation. Under the force of the impact, millions of tons of rocky debris from Earth's mantle were thrown into our planet's orbit. The debris ended up clinging together, forming a natural satellite that has a mass only a hundred times less than that of its parent planet. As for the metallic core of that falling planetoid, it would have fused with that of our planet.

The approximately 880 pounds (400 kg) of lunar rocks collected during the American *Apollo* missions, when 10 astronauts were able to dig in the lunar soil, played a crucial role in resolving this enigma. Dating of the samples indicates that our satellite formed 4.5 billion years ago, at the same time as our planet and the rest of the Solar System. The analysis revealed that, despite certain similarities (for example, the presence in both bodies of olivine, which is abundant in Earth's mantle), the chemical composition of terrestrial rocks differs from that of the lunar rocks. Planetologists have also solved other riddles about the moon, notably the origin of the "seas"—those dark expanses visible to the naked eye—and of the numerous craters marking the surface of the moon. The craters are the result of meteorite impacts, which occurred frequently at the beginning of the Solar System. The intense bombardment and the heat it released caused some portions of the crust and the underlying mantle to melt. Burning basaltic magma then poured out, filling the depressions and craters. When the bombardment stopped, the magma solidified. There is a connection between the presence of these seas and the fact that the Moon only ever turns one face toward us, the face that suffered the largest impacts and has the most extensive seas. The remains of those large asteroids, hidden in the subsoil, have created a surplus mass. This disequilibrium slowly brought the Moon to complete one rotation and one revolution around Earth during the same time period, about 27 days. Numerous other natural satellites also hide one side from their parent planet.

THE MOON IN NUMBERS

Average distance to Earth: 240,000 miles (384,000 km)

Diameter: 0.27 Earth diameters

Mass: 0.012 Earth masses

Temperature: −200°F (−130°C) in the sun; −275°F (−170°C) in the shade

Rotation: 27.3 Earth days

Revolution around Earth: 27.3 Earth days

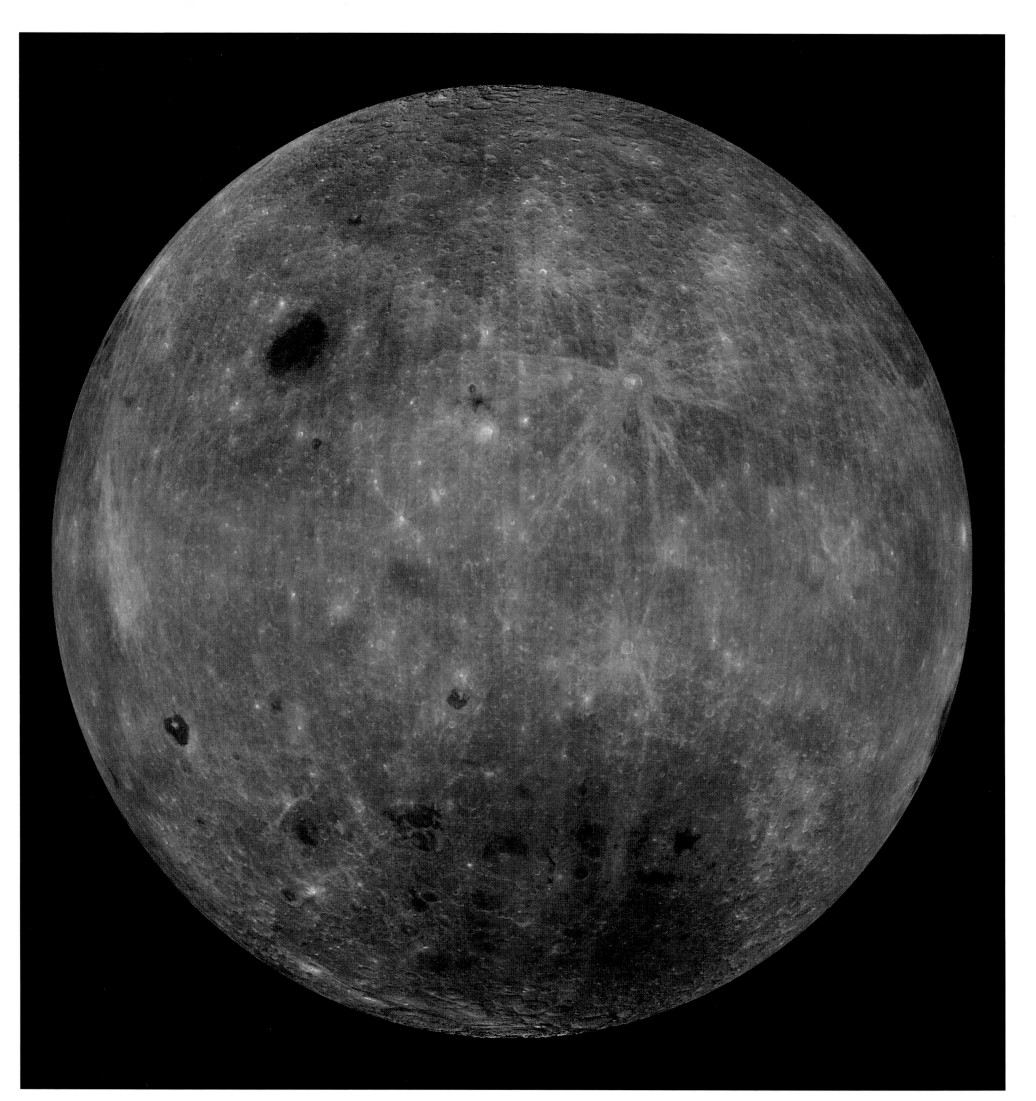

MERCURY
(MARINER 10)

Mercury, the closest planet to the Sun, is a desert world with a gray surface carved by solar winds and riddled by meteorite impacts of all sizes. Even though it is difficult to observe because of its proximity to the Sun, this planet has been known since prehistory. Much later, the Romans gave it the name of their messenger god with the winged feet.

No other object in the Solar System is subject to such dramatic temperature variations. At its zenith, the Sun heats the planet's surface to 800°F (430°C); at night the temperature drops to −300°F (−180°C). Both periods seem to last forever, as each day on Mercury lasts 58 Earth days.

The *Mariner 10* is the only spacecraft to have come close to this planet; it delivered the details of Mercury's morphology in some 12,000 images collected in 1974 and 1975. The image shown above is a mosaic of 18 shots taken at 42-second intervals on March 29, 1974, when the spacecraft was 125,000 miles (200,000 km) from the planet. It reveals the great abundance of craters, which often overlap. The smallest have microscopic dimensions, but many are larger than 125 miles (200 km) in diameter, such as the Kuiper crater, the most brilliant on the planet, which is visible near the center of the image. The most imposing, Caloris Planitia, attains a diameter of 750 miles (1,200 km). It was formed by the impact of an asteroid, and the seismic shockwave that followed it lifted mountainous rings over a mile (2 km) high all around the point of impact.

But Mercury was once the scene of a still more dramatic encounter. Shortly after its formation, a planetoid smashed into it, blowing part of the planet's rocky mantle into space and fusing with its metallic center. This is the current explanation for Mercury's enormous iron and nickel core, which takes up nearly 80 percent of the planet's radius.

MERCURY IN NUMBERS
Mass: 0.06 Earth masses
Diameter: 0.4 Earth diameters
Distance to the Sun: 0.36 AU
Temperature: from 880°F (470°C) with Sun at zenith to −280°F (−173°C) at night
Rotation: 58 Earth days
Revolution: 88 Earth days
Number of satellites: 0

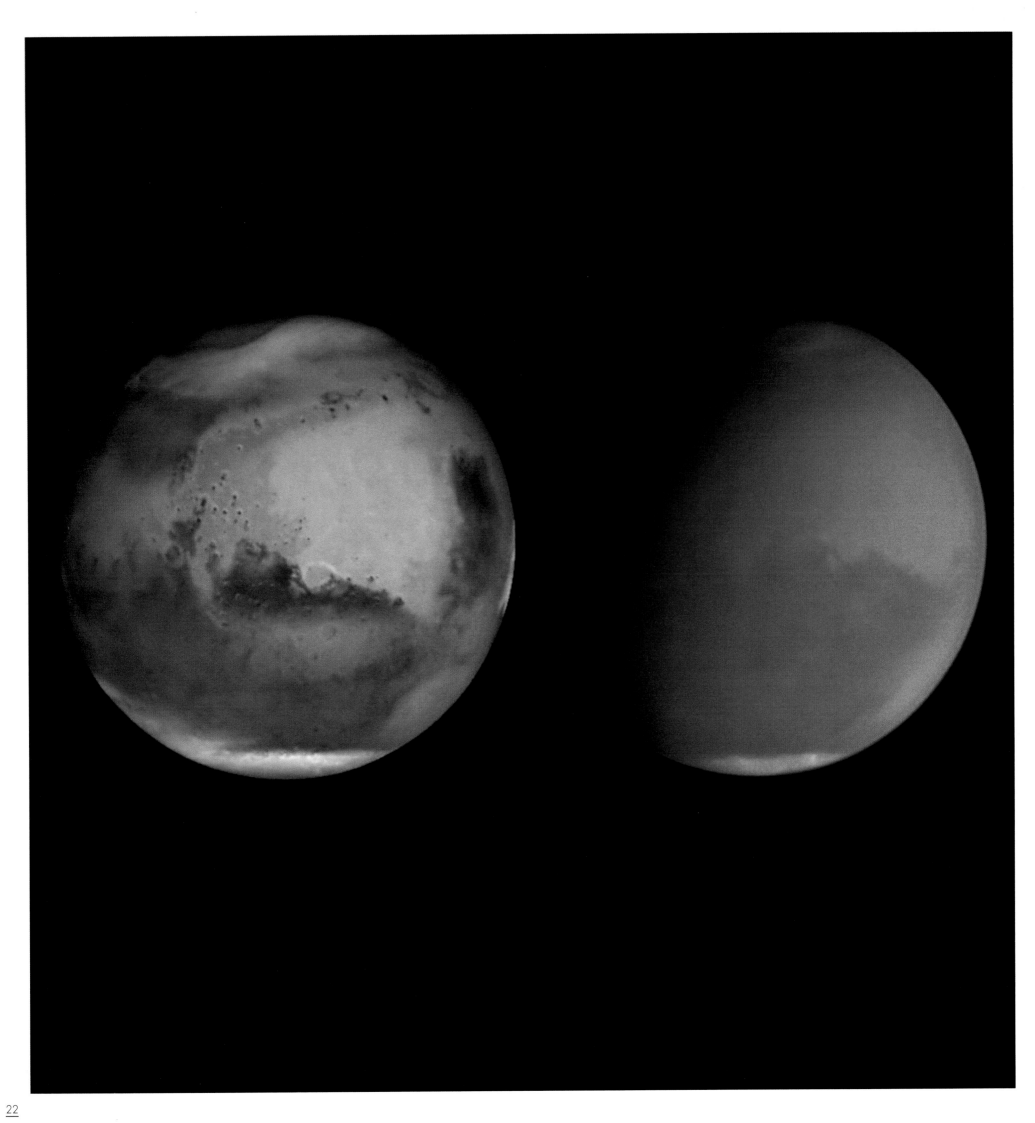

MARS BEFORE AND DURING A DUST STORM
(HUBBLE SPACE TELESCOPE)

The Hubble Space Telescope captured one of the most spectacular dust storms observed on the surface of Mars in recent decades. It all started with two innocent cyclones, visible in the photograph on the left, taken on June 26, 2001. One cloud formation is located in the north and the other, larger vortex whirls in the south, above Hellas Planitia (at lower right in the first image).

This weather is not surprising, because at the beginning of the Martian spring the ice of the polar caps—made up of carbon dioxide and water—partially sublimates, creating ice-crystal clouds. Further, the temperature difference between the polar caps and the sun-heated ground surface can reach 180°F (80°C), leading to winds of 125 miles per hour (200 kph). This time, a local meteorological event spread to the rest of the southern hemisphere. The disturbance then crossed the equator to start at least three other regional storms. A few weeks later, the entire planet was being swept by gales of red dust that continued for three months. At the height of the hurricane, on September 4, 2001 (photograph on the right), the veil of orange dust covered Mars almost completely. It was many long months before the atmosphere recovered its normal clarity

MARS
(VIKING 1)

This mosaic of images was taken in 1980 by the first spacecraft to land on Martian soil, *Viking 1*, and processed by computer to bring out the geological structures of Mars. To the west, three volcanoes—Arsia Mons, Pavonis Mons and Ascraeus Mons—stand out clearly in a line. Olympus Mons is located outside the frame, to the west of Ascraeus Mons. The dark stain that cuts into the north polar zone at the top right corresponds to Acidalia Planitia, a region of dunes formed by pulverized volcanic rocks.

Farther south can be seen the Valles Marineris canyon. This deep scar cutting through the Martian crust is 3,100 miles (5,000 km) long, 125 miles (200 km) wide and 5 miles (8 km) deep! Today we know that it formed during the cooling of the planet as the crust tore and settled, but until the end of the 19th century some astronomers thought that it was an irrigation canal constructed by inhabitants of the Red Planet. That thesis was passionately defended by the American Percival Lowell, who described some 160 canals.

Although the idea of a Martian civilization was discarded almost a century ago, numerous features similar to riverbeds are visible on the most recent pictures of the planet. One of them can be seen snaking through the center of this picture, linking the left part of Valles Marineris and the green structure farther north. These channels can extend up to 1,250 miles (2,000 km). Some close-up views of the canyon itself suggest that ravines were cut into it by water flows of catastrophic proportions, perhaps several times during the history of the planet—until the volcanoes became extinct, leading to the disappearance of liquid water 800 million years ago.

MARS IN NUMBERS
Mass: 0.107 Earth masses
Diameter: 0.53 Earth diameters
Average distance to Earth: 1.5 AU
Temperature: −72°F (−58°C) at the equator; −190°F (−123°C) at the poles
Rotation: 24 hours, 36 minutes
Revolution: 2 Earth years
Number of satellites: 2

JUPITER AND IO
(VOYAGER 1 AND HUBBLE SPACE TELESCOPE)

Jupiter, the enormous fifth planet, is twice as massive as all the other planets put together. These two images show clearly that its atmosphere is extremely turbulent. It is composed of gas, principally hydrogen and helium, mixed with traces of highly toxic compounds such as ammonia and hydrogen sulfide, which give it yellow-orange hues, and in some areas blue-green. The gaseous matter's arrangement into equatorial bands is linked to the planet's rotational speed. Jupiter completes an entire rotation in only 10 hours, which generates considerable centrifugal force and gives it a slightly oval shape, bulging at the equator and flattened at the poles. The gases follow this rotational movement, sometimes moving in the opposite direction, at different heights. Furious winds drive them. The darkest layers are the deepest and fastest, at 500 miles per hour (800 kph); the light-colored layers at the surface fly along at 185 miles per hour (300 kph). In some areas the gas is drawn into gigantic maelstroms, of which the most celebrated is the Great Red Spot. This cyclone (visible to the southwest in the first image) could swallow up several Earths. The light oval formations to the south appeared some 40 years ago.

Among other things, these two spectacular images reveal two of Jupiter's four Galilean satellites (named after their discoverer, Galileo). When the first picture was taken, Io was occulting its mother planet; in the second picture, Io is floating above Jupiter's disk, casting a shadow. Europa appears to the right in the first picture. Jupiter has over 14 other smaller satellites.

Like the other giant planets, Jupiter would have formed by progressively accumulating small chunks of rock and ice, which would have captured gases floating in the developing Solar System. At 33,000 feet (10,000 m) below the surface, the pressure exceeds one million times Earth's atmospheric pressure; there, hydrogen is a liquid.

JUPITER IN NUMBERS
Mass: 318 Earth masses
Diameter: 11 Earth diameters
Average distance to the Sun: 5.2 AU
Temperature: −220°F (−140°C)
Rotation: 10 hours
Revolution: 12 years
Number of satellites: over 16

IO SEEN IN FALSE COLOR
(GALILEO)

To measure the hold that Jupiter has on any object wandering too close to it, all it takes is a look at Io, the closest of the 16 Jovian satellites, which has a diameter close to that of our Moon. Io sustains a tearing internal friction that heats up its interior to the melting point. The resulting magma rises to the surface and spews out of the gaping mouths of numerous gigantic volcanoes. Rivers of lava snake for several hundreds of miles.

One element dominates the chemistry of this hell—sulfur, its color varying depending on the temperature and the chemical reactions it is undergoing. This explains the vivid colors of this body. They have been accentuated in this very high-definition image, which was obtained by selecting radiation emitted by Io in the near-infrared, green and purple portions of the spectrum. The data was acquired on July 3, 1999, when the *Galileo* spacecraft brushed past Io at a distance of 80,800 miles (130,000 km). Details of less than a mile (1.3 km) in length can be made out. The active volcanic zones betray their presence with shades of black, brown, green, orange and red. To the left of center, the fully active volcano Prometheus is letting fly an eruptive plume.

IO IN NUMBERS
Mass: 0.015 Earth masses
Diameter: 0.3 Earth diameters
Distance to the Sun: 5.2 AU
Distance to Jupiter: 0.025 AU
Temperature: −240°F (−150°C); 625°F (330°C) in volcanoes and lava lakes
Rotation: 1.8 Earth days
Revolution: 1.8 Earth days

POLAR AURORA ON JUPITER
(HUBBLE SPACE TELESCOPE)

On November 26, 1998, the space telescope captured a network of auroras illuminating Jupiter's polar atmosphere—the white glow against the vivid blue background. Such auroras, which also occur on our planet, arise when the solar wind penetrates a planet's magnetic shield near the poles, where it is weaker. The high-energy charged particles from the Sun ionize the atoms and molecules in the atmosphere, causing them to fluoresce in the form of vast, multihued curtains. In the case of Jupiter, however, the phenomenon proves to be more complex. The Galilean satellites cause the charged currents from the Sun to deviate, thus leaving their magnetic imprints next to the principal oval shape. The imprint of Io is on the left, that of Ganymede stands out in the center, inside the oval, and finally, Europa's imprint can be seen to the right, below Ganymede's.

THE RINGS OF JUPITER
(VOYAGER 2)

The *American Voyager 2* spacecraft took this picture on July 13, 1979, through orange and violet filters. It was 870,000 miles (1.4 million km) from Jupiter, and in its shadow (that is, the planet concealed the Sun).

The outermost layer of Jupiter's atmosphere stands out in backlight, along with surprising rings of dust. Their thickness is only 6 miles (10 km) for a length of 4,000 miles (6,500 km). The rings are the two straight orange segments that gleam weakly at the left. The lower segment appears to be interrupted because of the shadow cast by Jupiter on the ring. Are these rings the vestiges of an ancient moon, broken into a myriad of fragments by the titanic gravitational forces of the monstrous planet? No one knows. (The smearing of two of the structures in this image is caused by the movement of the spacecraft during the exposure.)

ICECAP COVERING EUROPA: CLOSE-UP VIEW OF THE MINOS LINEA REGION (GALILEO)

This image has been processed in false color to better reveal the surface of Europa. This smallest of the four Galilean satellites is one of the most fascinating worlds in the Solar System. The cameras on the *Voyager 2* spacecraft, which flew over it in 1979, revealed an ivory-colored body—probably covered with glacial ice—smoother than a billiard ball and streaked with thousands of sinuous interconnected lines only slightly darker than the rest of the surface. Some are only a few miles in length, while others extend for several hundred or even thousands of miles.

Nine years later, the American spacecraft *Galileo* was sent to Jupiter to examine the giant planet and its satellites in detail. In December 1995, after six years of travel, the craft finally reached its target. It performed several low-level flights above Europa and obtained numerous detailed images of the surface, including this one, recorded on June 28, 1996. The image covers a region about 785 miles (1,260 km) long. Glacial ice appears in blue and the shades of brown indicate three categories of structures— the *linea* are long, straight lines, the *flexus* are more sinuous and the *macula* are dark spots. The macula indicate a dark fluid, which may contain magnesium sulfates, rising up from within the interior. Europa seems to be covered by an immense fractured ice pack. Underneath it could be either a layer of dirty fluid ice or a real ocean of liquid water several dozen miles deep. Nothing can prevent us from speculating that underwater volcanic springs could be heating the water and enriching it with minerals. Who knows whether oases teeming with life might be found in the depths of Europa!

EUROPA IN NUMBERS
Mass: 0.08 Earth masses
Diameter: 0.25 Earth diameters
Distance to the Sun: 5.2 AU
Distance to Jupiter: 0.05 AU
Temperature: −260°F (−163°C) at the equator; −370°F (−223°C) at the poles
Rotation: 3.5 Earth days
Revolution: 3.5 Earth days

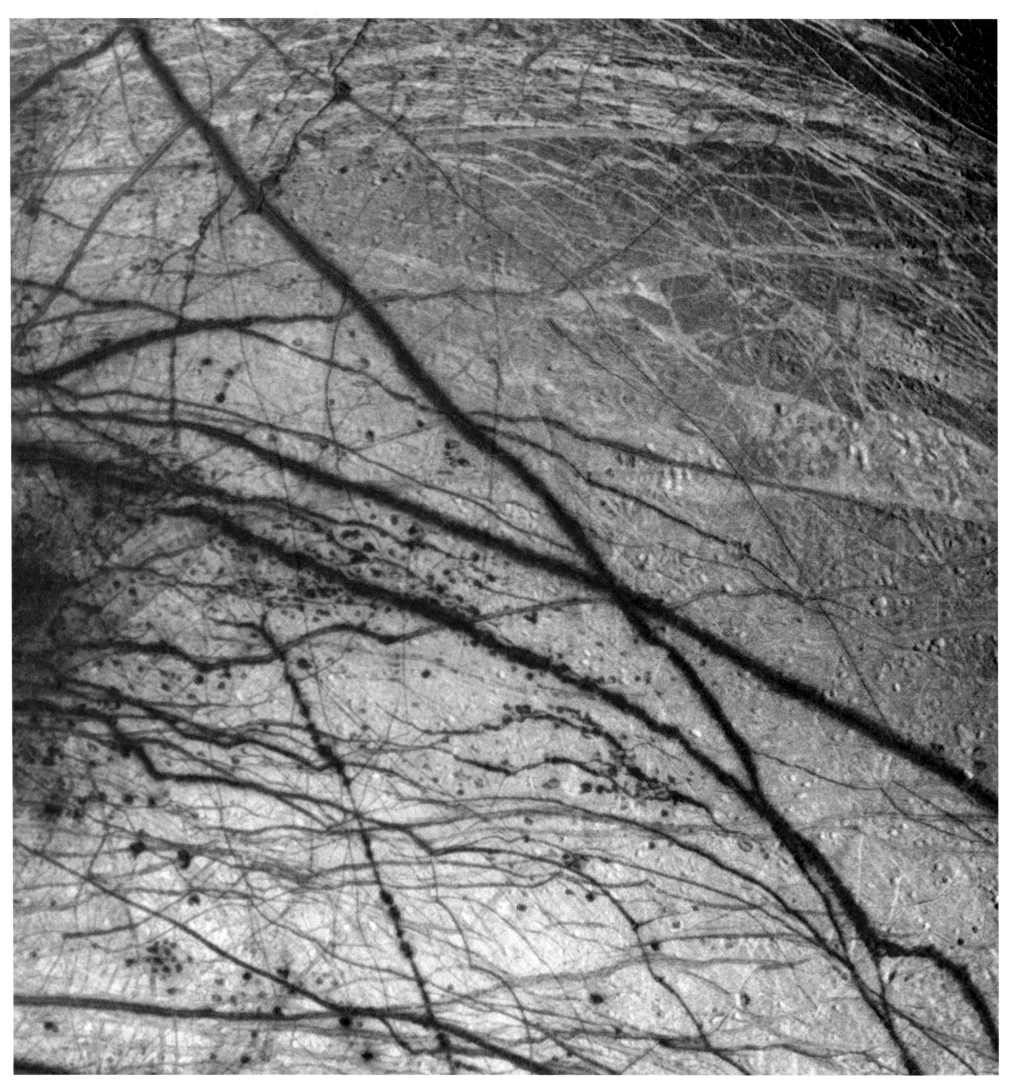

SATURN AND ITS RINGS
(VOYAGER 2)

Instantly recognizable among the planets because of its brilliant rings, Saturn, the sixth planet, provides a grandiose spectacle. Saturn is the second-largest planet in the Solar System, after Jupiter. The two have similar compositions, with a small rocky core covered by layers of liquid hydrogen, and then a gaseous envelope made turbulent by violent atmospheric currents and titanic hurricanes. All the same, there is less contrast between these structures on Saturn than on Jupiter.

The first photo is presented in realistic color and the one to its right in enhanced color to increase the contrast between the planet's cloudy bands. Three satellites of Saturn—Tethys, Dione and Rhea—stand out against the black background. Mimas, a fourth, smaller moon, is visible as a light-colored dot against the cloud cover on the left, below the rings. The silhouettes of Mimas and Tethys are projected as shadow patterns on the giant planet's disk.

The rings were discovered by the Dutch astronomer Christiaan Huygens in 1656. Their diameter reaches 186,000 miles (300,000 km), but they scarcely exceed a few hundred yards in thickness. The rings contain myriad ice-covered rocks, ranging from microscopic to several yards across. At high resolution, the rings are seen to be made up of a multitude of smaller rings, some brighter than others, undulating in some areas as if propelled by currents. Several bands containing little material, called divisions, separate them. In the two color images the most important of these can easily be seen—Cassini's Division, between Ring B on the inside and Ring A on the outside.

The *Voyager 2* spacecraft took the close-up view on August 22, 1981, when it was 2.5 million miles (4 million km) from the giant planet. Scientists observed enigmatic formations, probably constantly changing—the "strokes," seen as black marks in the rings at the center of the image. In these zones, microscopic dust may be being levitated above or below the plane of the rings. A magnetic storm caused by ionized particles from the solar wind could be responsible for this phenomenon.

The precise origin of Saturn's rings remains to be discovered, and two hypotheses are under discussion. According to the first one, the rings may be unconsolidated residues of materials—dust grains and small rocky and icy bodies—that were present in the disk from which the Solar System formed. The second hypothesis holds that they result from one (or maybe several) former satellites that orbited too close to Saturn, where the tidal forces produced by the planet caused their disintegration.

SATURN IN NUMBERS
Mass: 95 Earth masses
Diameter: 9 Earth diameters
Distance to the Sun: 9.5 AU
Temperature: −288°F (−178°C) at the surface
Rotation: 10.2 hours
Revolution: 29.5 Earth years
Number of satellites: 18

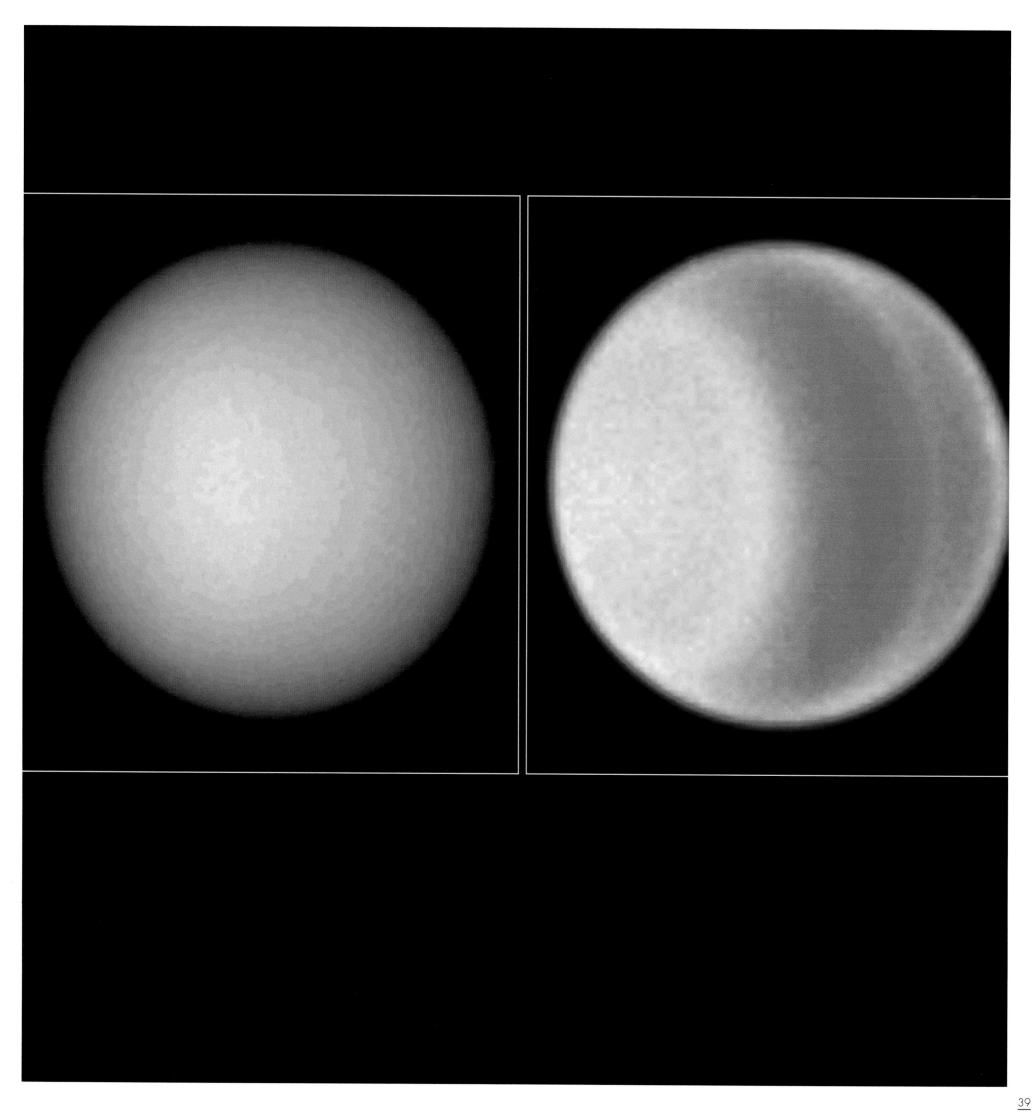

PAGE 38

CRESCENT URANUS
(VOYAGER 2)

PAGE 39

URANUS: NORTHERN WINDS
(HUBBLE SPACE TELESCOPE)

Uranus, the seventh planet of the Solar System, is a gaseous planet whose atmosphere gives off a blue-green color; methane (natural gas), which it contains in very low quantities, absorbs red and yellow light. The American *Voyager 2* spacecraft took the backlit view of its dense atmosphere in January 1986, when it was pulling away from the planet. The photo reveals neither nuances of color nor possible atmospheric structures.

Eleven years later, the Hubble Space Telescope detected for the first time the presence of clouds in Uranus's northern hemisphere. The green image is a black-and-white view that has been colored by computer to mimic the natural color of the planet. The globe seems uniform, except for the presence of a small cloud on the right edge. The red-and-pink image shows absorption of light by the molecules of methane in the atmosphere. In addition to the cloud just described, it reveals discrete cloudy bands circling the planet.

Observers hope that these structures will separate more and more sharply as Uranus leaves its winter season and warms up. The year on Uranus is 84 times longer than on Earth. Also, the planet is completely tilted over in its orbit; the north pole is thus in total darkness for nearly half a century.

URANUS IN NUMBERS
Mass: 14.5 Earth masses
Diameter: 3.9 Earth diameters
Distance to the Sun: 19.2 AU
Temperature: −365°F (−220°C) at top of atmosphere; 36,000–54,000°F (20,000–30,000°C) at center
Rotation: 17.2 hours
Revolution: 84 Earth years
Number of satellites: 17

PAGE 41

NEPTUNE
(VOYAGER 2)

Neptune, the eighth planet, was discovered in 1846 at a position in the sky that had been predicted by theoretical calculations. It is the twin sister of its neighbor Uranus, and both are essentially gaseous. Its thin atmosphere, rich in hydrogen and helium, covers a liquid mantle of water, methane and ammonia that is subjected to Herculean pressures (more than 100,000 times Earth's atmospheric pressure) and raised to several thousand degrees. The center of the planet may be made up of a rocky core.

The surface of Neptune was revealed by the *Voyager 2* spacecraft, which took this picture in 1989, when it was below the planet and in its shadow. Neptune is the second "blue planet" in the Solar System, after Earth, but do not be fooled by the color. It is in no way due to the presence of oceans of liquid water, but comes from clouds of methane ice crystals floating high in the atmosphere. This atmosphere seems more complex and dynamic than that of Uranus. *Voyager 2* revealed the presence of an anticyclonic dark spot, similar to the Great Red Spot on Jupiter, where winds rage at over 600 miles per hour (950 kph).

Neptune remains the most distant celestial object ever explored by a spacecraft. *Voyager 2*, like *Voyager 1*, has since left the Solar System and is currently heading into interstellar space. They are truly messages in bottles in the ocean of space, destined for possible extraterrestrial civilizations—both carry videodiscs containing information about our world. Only after 80,000 years, or even longer, will they be approaching other stars.

NEPTUNE IN NUMBERS
Mass: 17.2 Earth masses
Diameter: 3.8 Earth diameters
Distance to the Sun: 30 AU
Temperature: −355°F (−215°C) at top of atmosphere; 36,000°F to 54,000°F (20,000°C to 30,000°C) at center
Rotation: 16.22 hours
Revolution: 165 Earth years
Number of satellites: 8

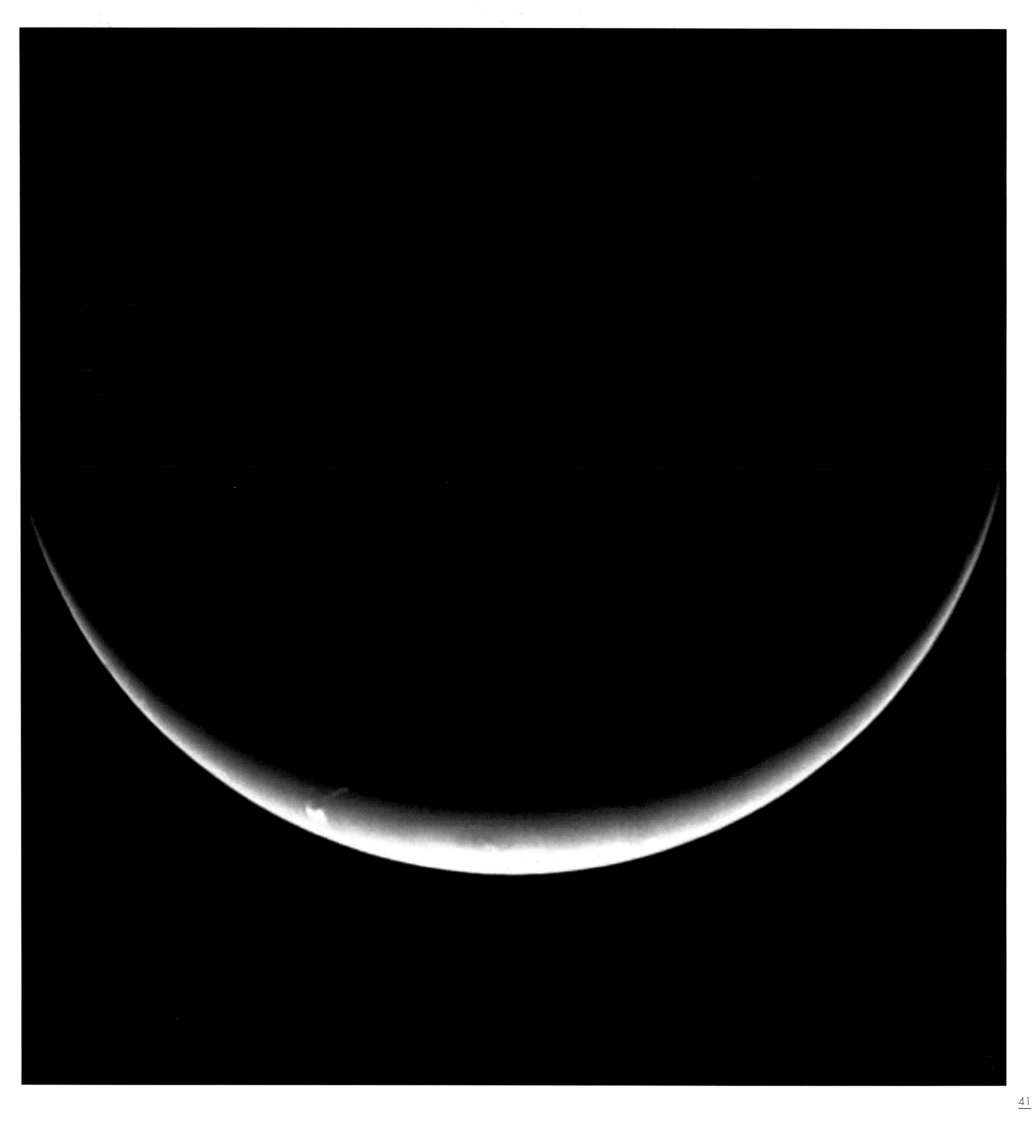

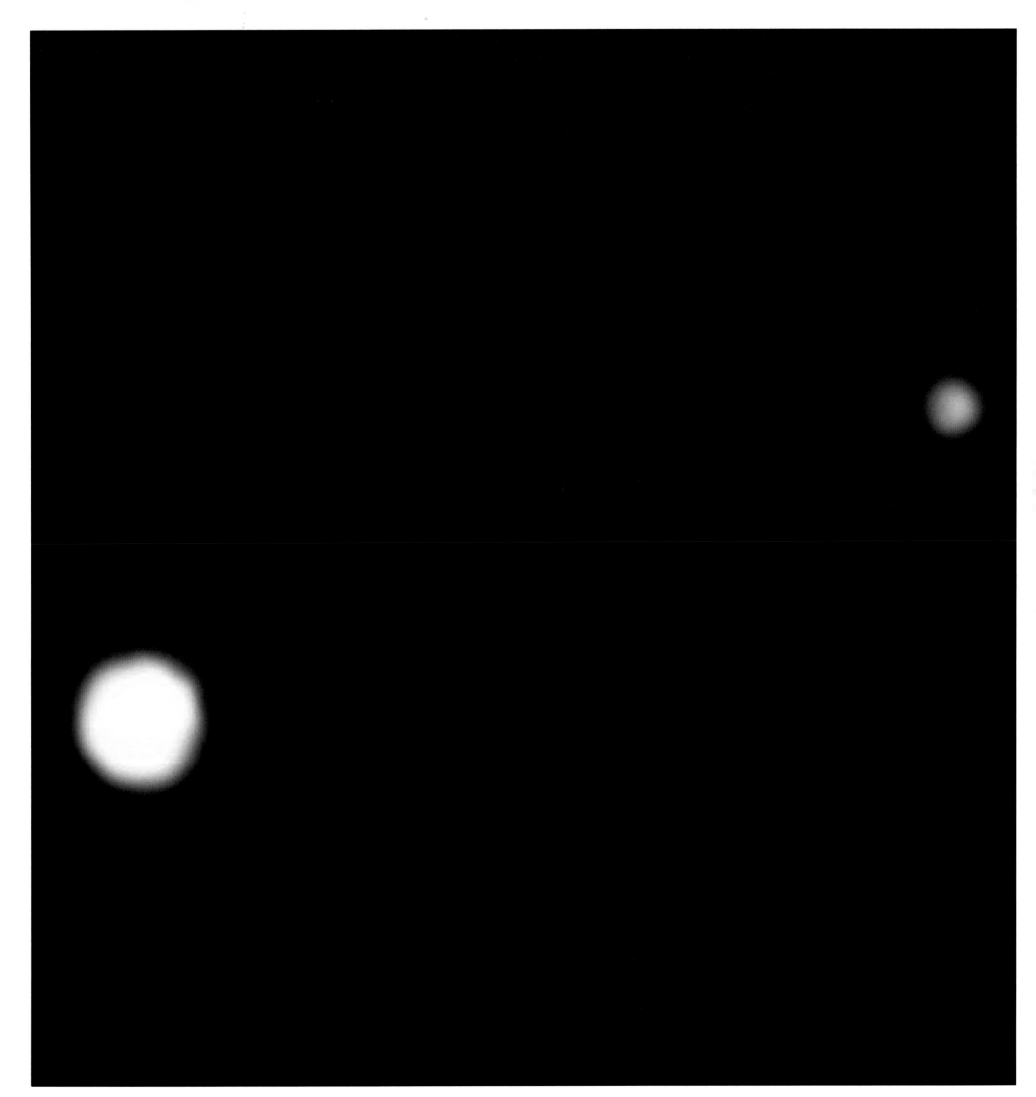

Pluto and its satellite, Charon

Pluto, the ninth planet, is the smallest and most remote in the Solar System, and has not yet received a visit from any spacecraft. It is thought to be a rocky, bitterly cold world; frozen water, methane, nitrogen, ammonia and carbon monoxide have been detected on it. Its satellite, Charon, was discovered only in 1978. It is much closer to its planet than such satellites tend to be. Seen from the surface of Pluto, it seems never to move in the sky, and its apparent size is over four times that of the Moon seen from Earth.

During the summer of 1994, the Hubble Space Telescope sent us the first detailed information obtained about the surface of Pluto. However, many aspects of Pluto remain mysterious. Why is this planet tilted in its orbit? Did it undergo a collision? Is it a former satellite of Neptune that escaped from the gravitational hold of its parent planet? Or is it a large asteroid? Pluto does indeed orbit at the inner edge of the Kuiper Belt, a disk of frozen objects of all sizes, from large asteroids to dust grains. NASA is preparing an exploratory mission to unveil these mysteries. The spacecraft *New Horizon* should take off from Florida in 2006 and reach Pluto, and then the Kuiper Belt, more than nine years after its launch.

PLUTO IN NUMBERS
Mass: 0.02 Earth masses
Diameter: 0.20 Earth diameters
Distance to the Sun: 40 AU
Temperature: −380°F (−230°C)
Rotation: 6 Earth days
Revolution: 248 Earth years
Number of Satellites: 1

44

COMET HALE-BOPP

Visible for several weeks in 1997, the comet Hale-Bopp is perhaps the most beautiful ever observed. It is one of billions of small bodies that have been preserved unchanged since the beginning of the Solar System 4.5 billion years ago. In our epoch they appear as part of a vast disk made up of fragments of rock and ice. Earlier, some of them fused to form planets and their satellites, and the leftovers were condemned to wander in solitude. Thousands bombarded the planets as they were being formed or were consumed in the Sun. The rest collected in the most remote parts of the Solar System—the Kuiper Belt, which extends beyond Pluto up to 500 AU from the Sun, and the even larger Oort Cloud, which reaches out to 100,000 AU. The latter is a reservoir of billions of dirty snowballs about 6 miles (10 km) in diameter.

Subject to gravitational disturbances and even collisions with other bodies, some of these fragments may be hurled toward the Sun and enter an orbit around it. As the ball of ice approaches the Sun and its nucleus heats up, the ice vaporizes and it begins to drag dust along behind it. The dust spreads out in a white or yellowish tail that reflects the light of the Sun. Hale-Bopp's second bright blue tail contains cometary gases being expelled into interplanetary space. Ultraviolet (UV) radiation from the Sun causes the gases to ionize and consequently to fluoresce; the solar wind pushes this tail away from the Sun. Unfortunately, Comet Hale-Bopp has a very elliptical orbit, so it will not be back for another 2,400 years

Leonid meteor shower

It is a dazzling spectacle—in one hour dozens of bright streaks pass before you. They are meteors, and they all seem to be shooting from the same point in the sky, which is called the radiant. Showers of shooting stars occur when Earth crosses the orbit of a meteoroid swarm composed of dusty debris expelled by a comet during its passage near the Sun. The swarm follows a path similar to that of the associated comet, and the grains of dust vaporize when swept into Earth's atmosphere.

The shower shown in this picture comes from Comet Tempel-Turtle, whose period of revolution is 33.2 years. This picture, which needed three 20-minute exposures, was taken in November 2001 during the height of the swarm's annual activity. The comet's peak is around November 17, but usually the shower that goes with it is quite subdued. However, when the comet returns every 33 years, spectacular showers are produced. In 1966 over a thousand meteors an hour shot across the sky. This phenomenon was repeated in 1999, as well as in the two following years.

The name of the shower comes from the constellation in which its radiant is located—Leo, the Lion. Other beautiful meteor showers are the Taurids in June/July, the Perseids in August, the Orionids in October and the Geminids and Ursids in December.

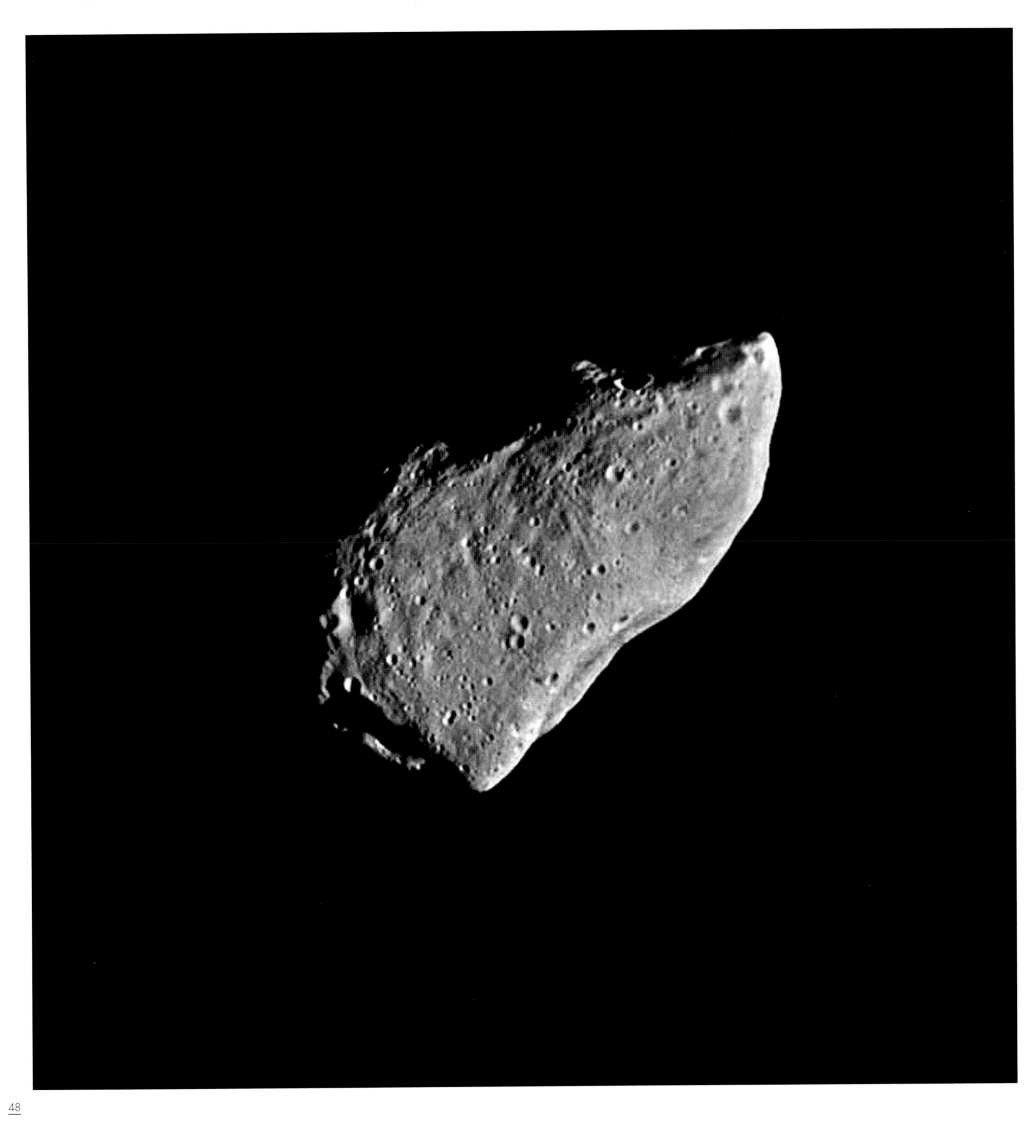

PAGE 48

ASTEROID 951, GASPRA
(GALILEO)

This pebble in the sky is an elongated asteroid discovered in 1916. It circles at 205 million miles (330 million km) from the Sun among the multitude of similar fragments that make up the principal asteroid belt, located between Mars and Jupiter.

Gaspra is the first asteroid that a spacecraft has ever flown by. In October 1991 the *Galileo*, on its way to Jupiter, passed within about 3,300 miles (5,300 km) to take 57 pictures, revealing for the first time specific information about the morphology and surface of these previously enigmatic objects (seen through telescopes they are simply white spots).

Gaspra's precise measurements (11 by 7 by 5.5 miles / 18 by 11 by 8.9 km), its mass (5 trillion tons) and the appearance of its surface are now known. It is covered by a fine dust of broken rock, or regolith, and its surface is extremely uneven. About six hundred impact craters prove that it has undergone numerous collisions with other asteroids—fewer, however, than for objects formed at the beginning of the Solar System, as if Gaspra is much younger. It is possible that it could be part of the debris of an enormous older asteroid, itself a conglomeration of rocky chunks, that broke up during a collision only 200 million years ago.

PAGE 49

ASTEROID 243, IDA, AND ITS SATELLITE, DACTYL
(GALILEO)

Ida is the second asteroid that a spacecraft has flown by. It is very elongated (32 by 15 by 12 miles / 52 by 24 by 20 km). This uneven block of rock was photographed on August 28, 1993, from about 1,550 miles (2,500 km) by the CCD camera on *Galileo*, while it was en route to Jupiter. The images revealed a surprise: Ida was accompanied by a mini-asteroid of at least 1.25 miles (2 km) in diameter—given the name Dactyl—that orbits it at a distance of about 60 miles (100 km). In the photograph, Dactyl (on the right) is closer to the space vehicle than Ida. Since that discovery, astronomers have discovered a mini-satellite around another asteroid, Eugenia (discovered in the 19th century) that is over 125 miles (200 km) in diameter. The bluish zones around the craters at the upper left, center and upper right reveal an abundance of compounds rich in iron. Ida, its moon and several other neighboring asteroids are members of the same family, Koronis. They probably all come from a massive body fragmented by a violent collision.

PAGE 51

THE EARTH-CROSSING ASTEROID EROS
(NEAR)

In 1996 the NEAR (Near-Earth Asteroid Rendezvous) spacecraft was sent off to the asteroid Eros. Four years later the small craft hovered barely 6 miles (10 km) above the asteroid's surface, not landing on it, but revealing totally unexpected and detailed information. Eros's measurements are 25 by 9 by 9 miles (40 by 14 by 14 km) and it weighs 6.7 billion tons. Unlike most asteroids, 95 percent of which normally never go beyond the limits of the main asteroid belt between Mars and Jupiter, Eros is part of a family of a hundred wandering rocks whose orbits cross that of Earth. They are called Earth-crossers or NEOs (for Near-Earth Objects). These Earth-crossers—Apollo, Adonis, Hermes, Icarus, Toutatis, Sisyphus, etc.—are now under constant surveillance. After all, the risk of a collision with Earth is not negligible. Our planet has already sustained more than 150 large impacts, including that of a bolide, or meteor, 6 miles (10 km) in diameter. That struck Mexico 70 million years ago, causing a genuine climatic disaster. The dust that it threw into the atmosphere plunged Earth into darkness for several years, and most living species, including the dinosaurs, became extinct.

The destruction it creates is a function of an asteroid's size. A bolide 250 feet (75 m) in diameter would wipe a large city from the map. A meteor one mile (1.7 km) in diameter could destroy France and disrupt the climate of the entire planet, notably by making the ozone layer, which protects us from harmful ultraviolet rays, disappear. An event of such magnitude theoretically occurs, on average, every 250,000 years.

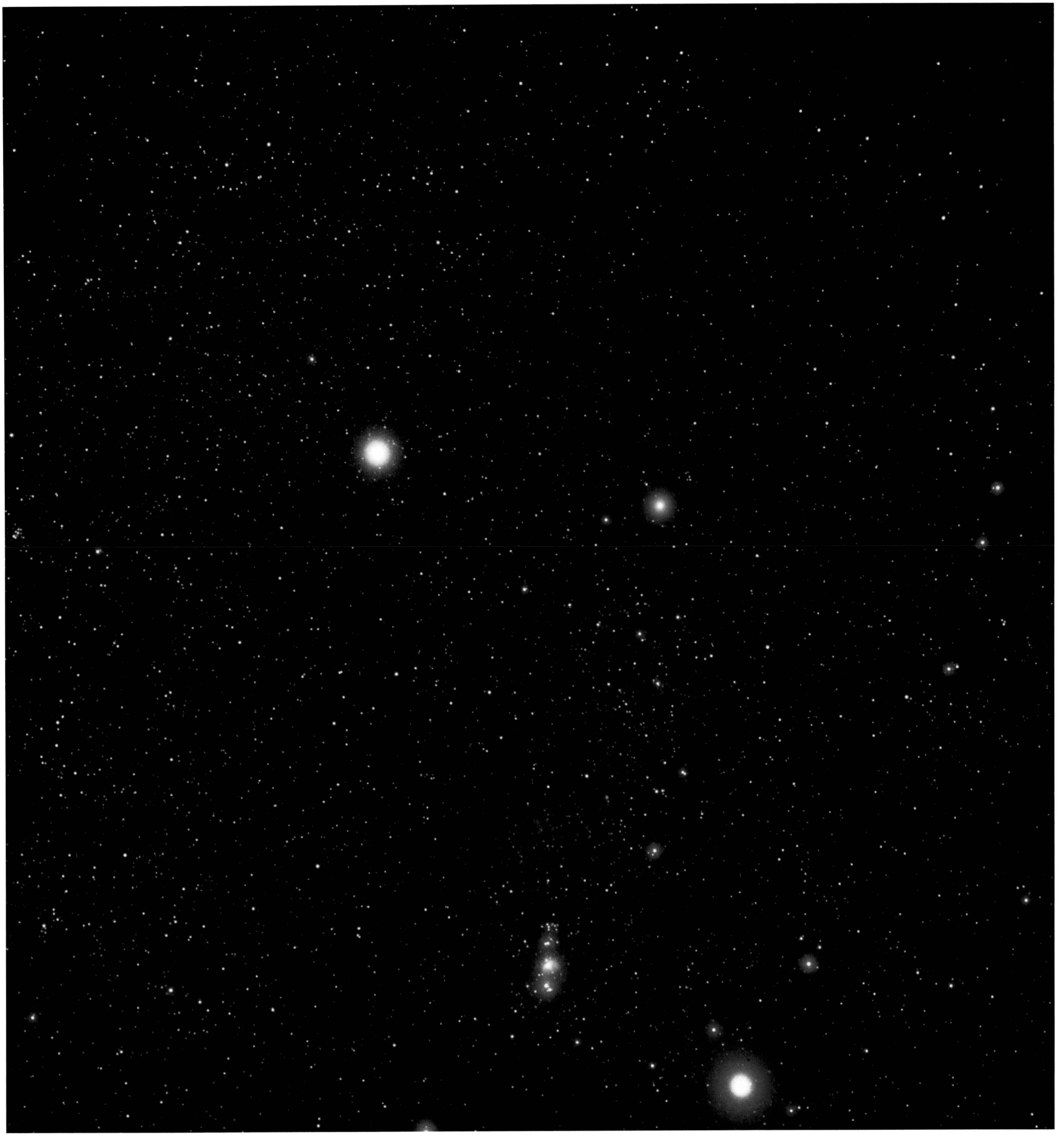

DIFFUSE NEBULAS AND STAR BIRTH

Interstellar space is not at all empty. Immense clouds of gas and dust spread out between the stars. They are made up in part of original matter that came from the very beginning of our universe, 15 billion years ago, and in part of gas and dust expelled into the cosmos by the death throes of stars.

Some of these dark, thick clouds form shadows against the cosmic background, hiding stars and other hazy bodies. Other nebulas simply reflect the light of nearby stars. Still others display wonderful crimson draperies whose beauty only certain photographic or electronic media can manage to record. To the naked eye they appear grayish, because in low light the retina is incapable of perceiving colors.

These nebulas are the cosmic stuff from which new stars are generated. We have known for just a little while now that they break up into enormous "cocoons," each containing a budding star. And so, enriched by the nourishing ash of decomposed stars, like humus in the soil, these nebulas participate in a gigantic cycle of life and death on a cosmic scale.

The constellation of Orion

PAGE 52

CONSTELLATION OF ORION

On a clear night in January, around midnight, dress warmly and go outside armed with binoculars. Looking toward the south, you will be able to admire one of the most beautiful constellations in the sky—Orion. The constellation was named by the Greeks in honor of the giant hunter, described by Homer, who was stung by a scorpion.

Orion contains some remarkable stars. The orange blaze at top left is the star Betelgeuse, 400 light-years from the Sun, shining more brightly than 10,000 Suns. It is a star on the decline—pulsating, and already in the red supergiant stage. Its color contrasts sharply with the blue brilliance of the other stars shown here: the blue supergiants Meissa (top), Bellatrix (top left) and even Rigel (lower left), whose name means "foot" in Arabic. The "Three Kings" aligned in the center are named Alnitak, Alnilam and Mintaka.

Why these different colors? The color of a star gives information about its surface temperature and also its mass, size and life expectancy. In fact it glows like a piece of heated iron, which turns progressively red, orange, yellow, white and, finally, blue.

This constellation also contains magnificent diffuse nebulas (shown in more detail on the following pages), of which the brightest in the sky—four times more intense than the full Moon—is the celebrated Messier 42, its pinkish glow visible below the Three Kings. The blue glow above it comes from a smaller nebula reflecting its light. Just below the first King is the Horsehead Nebula, which is too faint to be visible in this image. An even fainter nebula, in the shape of a witch's head, floats to the right of Rigel. Together these objects make up part of a gigantic cloud of gas and dust that travels this region of the sky.

PAGES 54 AND 55

ORION NEBULA, MESSIER 42
(VERY LARGE TELESCOPE)

Until 1610 the Orion Nebula was thought to be a star; it is in fact a diffuse nebula. It is also the closest nebula to Earth (1,200 light-years away) and one of the most active celestial incubators. Several tens of thousands of stars, today dispersed far and wide, were probably born there.

The youngest, from a hundred thousand to one million years old, have not yet left their nest; they are the four stars of the Trapezium, located at the center of the nebula. The intense stellar wind that they give off has blown away the gas, giving the nebula the appearance of a shell. What's more, they heat the ambient gas to over 18,000°F (10,000°C), ionizing the elements present—notably hydrogen, which fluoresces red like a neon light.

The second photograph shows an enlarged view of the heart of the nebula, from which only infrared radiation has been captured. It shows the presence, in the area of the Trapezium, of a myriad of newborn stars concentrated in an area scarcely larger than the distance between the Sun and Proxima Centauri, the star closest to Earth.

Images at an even higher resolution, obtained since 1992, have shown that some 50 of these stars are surrounded by dark clouds of a size comparable to that of our Solar System. These clouds are likely made up of matter orbiting the stars, quite possibly fusing to form planets. For that reason they have been named proto-planetary disks, or proplyds.

TECHNICAL DATA (IMAGE 2)
Mosaic of 81 images taken by the ISAAC infrared camera on the Very Large Telescope, Chile, December 20 and 21, 1999. The field of the photograph measures 3 by 3 light-years.

PAGE 57

DETAIL OF THE HORSEHEAD NEBULA, BARNARD 33
(VERY LARGE TELESCOPE)

This nebula in the constellation Orion was photographed for the first time in 1888, by the observatory of Harvard University. Since then it has become one of the most admired objects in the sky. An enormous opaque molecular cloud stands out like a shadow in front of the pinkish halo of an emission nebula lit by the star Zeta Orionis, or Alnitak (located outside the field of view). The horsehead shape is actually the upper extension of a much larger cloud (below the frame). The scrolls and curves of this outgrowth are tangled in a complex manner because of their turbulent motions. The blue-green colors are caused by light from neighboring stars reflected on the dusty curls that fringe the shape.

The stellar winds and ionized gases of this colored nebula are gradually eroding its magnificent scrolls, so the head will gradually lose its familiar shape over the course of thousands of years.

TECHNICAL DATA
This image results from the superimposition of three pictures obtained February 1, 2000, using the 27-foot-diameter (8.2 m) KUEYN instrument of the Very Large Telescope, European Southern Observatory, Chile. Filters used: B-band (429 nm; FWHM 88 nm) rendered in blue; V-band (554 nm; 112 nm) in green; R-band (655 nm; 165 nm) in red.

WITCH HEAD NEBULA, IC 2118

Hooked nose, toothless mouth and jutting chin—here is the witch of the sky. Even though it occupies a portion of the sky as large as six full Moons, this nebula is too faint to be visible to the unaided eye. It is a cloud of microscopic dust located at the foot of the Orion Nebula, in the direction of Eridanus. It reflects light from the star Rigel, or Beta Orionis (outside the frame of this picture), a blue supergiant that is seventh on the list of stars most brilliant to the naked eye—it shines like 40,000 Suns.

However, the cloud appears bluer than the light from the star itself. This is because the dust particles do not reflect incident light uniformly. Like microscopic particles suspended in Earth's atmosphere or in cigarette smoke, they reflect more blue radiation because they absorb blue less than the other colors.

TECHNICAL DATA

The photograph was taken from Joshua Tree National Park in California, January 20, 2001, using a Takahashi FSQ-106N telescope with four fluorite elements at f/5 and a CCD camera (1530 x 1020 pixels / 9 x 9 mm). Twenty-two exposures were made. Total duration: 3.7 hours. Filters: blue, red and green.

PAGE 60

CONSTELLATION OF SCORPIUS AND PIPE NEBULA
(DAVID MALIN)

The scorpion that, according to legend, poisoned the giant Orion has its place on the other side of the sky, within a very beautiful celestial region visible from the Southern Hemisphere. The astrophotographer David Malin obtained this panorama, as well as the next view, centered on the arachnid's abdomen. He captured colors that are both vivid and realistic, but imperceptible to the naked eye because "at night, all cats are gray"—the cells of the retina that are sensitive to colors cannot function when the light is too dim.

Immense filaments of dusty gas can be distinguished, obscuring the field of stars. One of them has the shape of a pipe (in the middle of the photo on the left side). Below its stem (at the bottom edge of the frame) is the emission nebula NGC 6357.

PAGE 61

CONSTELLATION OF SCORPIUS, REGION OF ANTARES AND RHO OPHIUCHI
(DAVID MALIN)

The arachnid's abdomen contains a very beautiful collection of colored clouds. The yellow star (on the right in the previous image and at lower left in this photograph) is Antares—meaning "anti-Mars"—a red supergiant that occupies 16th place on the list of stars most brilliant to the naked eye. Near death, it is weaving its own burial shroud, solid particles that reflect its light. Sigma Scorpii, farther right, is a giant star that is partly hidden by the diaphanous veils of a red emission nebula. At the top of the picture, a dust-cluttered layer conceals the background stars and reflects light from the star Rho Ophiuchi.

Another remarkable body is present here—the globular cluster Messier 4 (lower right), which is 8,000 light-years from the Solar System

PAGE 64

DARK CLOUD BARNARD 68 IN THE CENTRAL REGION OF THE MILKY WAY
(VERY LARGE TELESCOPE)

PAGE 65

DARK CLOUD BARNARD 86 IN THE CENTRAL REGION OF THE MILKY WAY

Holes in space-time? Tunnels to other worlds? The answer is no, and much more prosaic—the dark zones in these two pictures are large clouds of gas and dust floating in the telescope's line of sight. They cut into the very dense fields of stars in the southern constellations Serpens (first picture) and Sagittarius (second picture).

These molecular clouds are called Bok globules or Barnard objects, after the scientists who discovered and cataloged them. The physical conditions inside these clouds are becoming better and better understood. Their density can reach one million times that of interstellar space, which is one atom per cubic centimeter, on average. But that is still very much less than Earth's atmosphere, which can reach several hundreds of billions of billions of atoms per cubic centimeter. Within the clouds, temperatures scarcely exceed –440°F (–263°C), which makes them the coldest objects in the universe. A hundred kinds of molecules have been detected inside them, including water, methane (natural gas), ethanol, formic acid (which ants use to mark their paths!) and amino acids (the basis of life on Earth), along with large structures of organic molecules.

Some molecules fuse to form dust grains. The smallest among them—0.002 micron in diameter (a micron is one-millionth of a meter)—seem to be polycyclic hydrocarbons, similar to those in automobile exhaust. The largest are silicates—materials based on silicon, magnesium and iron that are found in terrestrial rocks— and can reach a micron in diameter, the size of particles in cigarette smoke. They are undoubtedly covered with a layer of ice composed of water, methane or ammonia.

These clouds seem to come from larger clouds that have fragmented. The data shows that some of them, notably the one in the first picture, which was obtained in 1999 using the Very Large Telescope, are undergoing gravitational collapse. This process could continue for several hundreds of thousands of years to form "cocoons" sheltering stellar embryos, as in the Orion and Eagle nebulas (see pages 54 and 55 and pages 86 and 87).

The cloud in the second photograph is more mysterious. Will it too become a stellar nursery? Or is it a relic of a gaseous nebula, once richly colored, that gave birth to the other curiosity in this picture—the young stars in the open cluster NGC 6520 beside it?

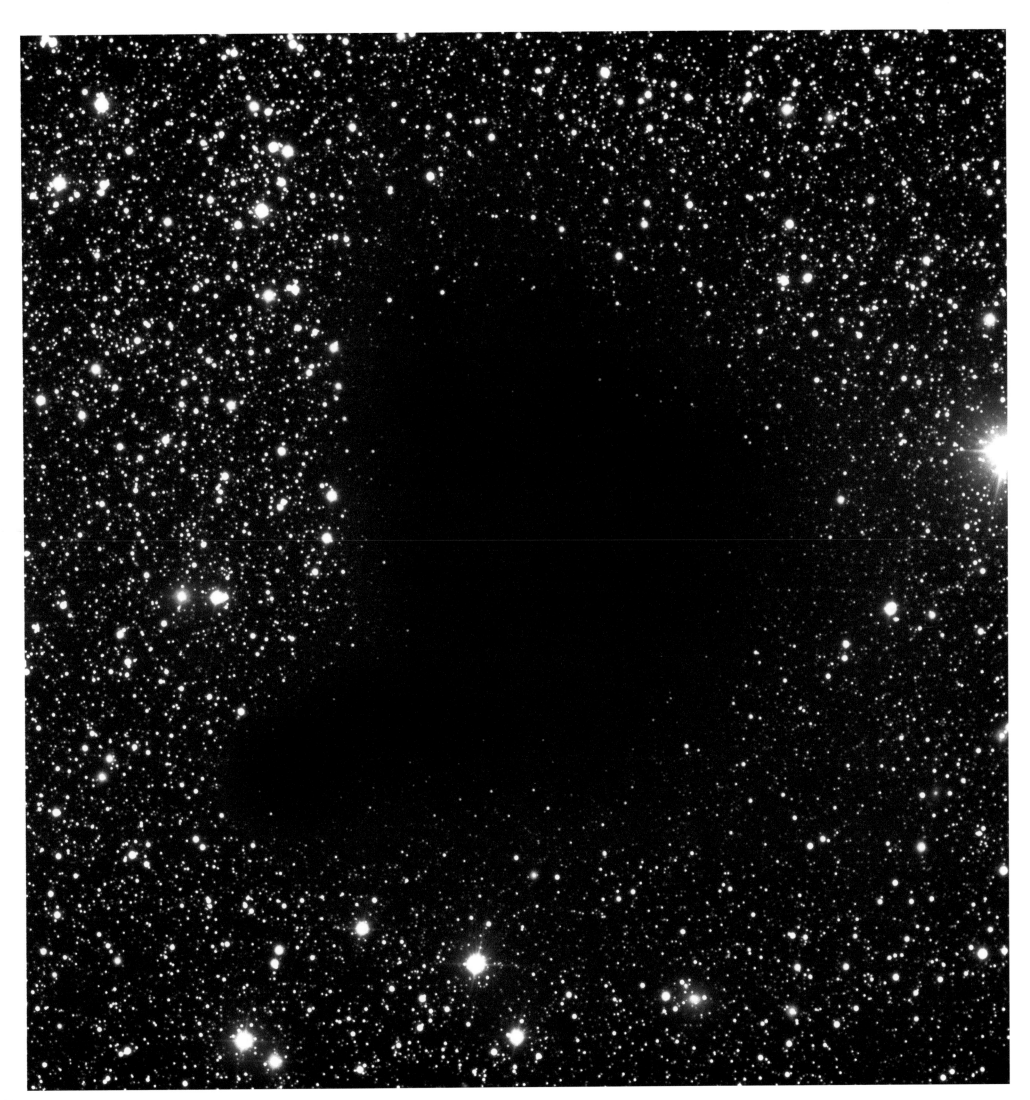

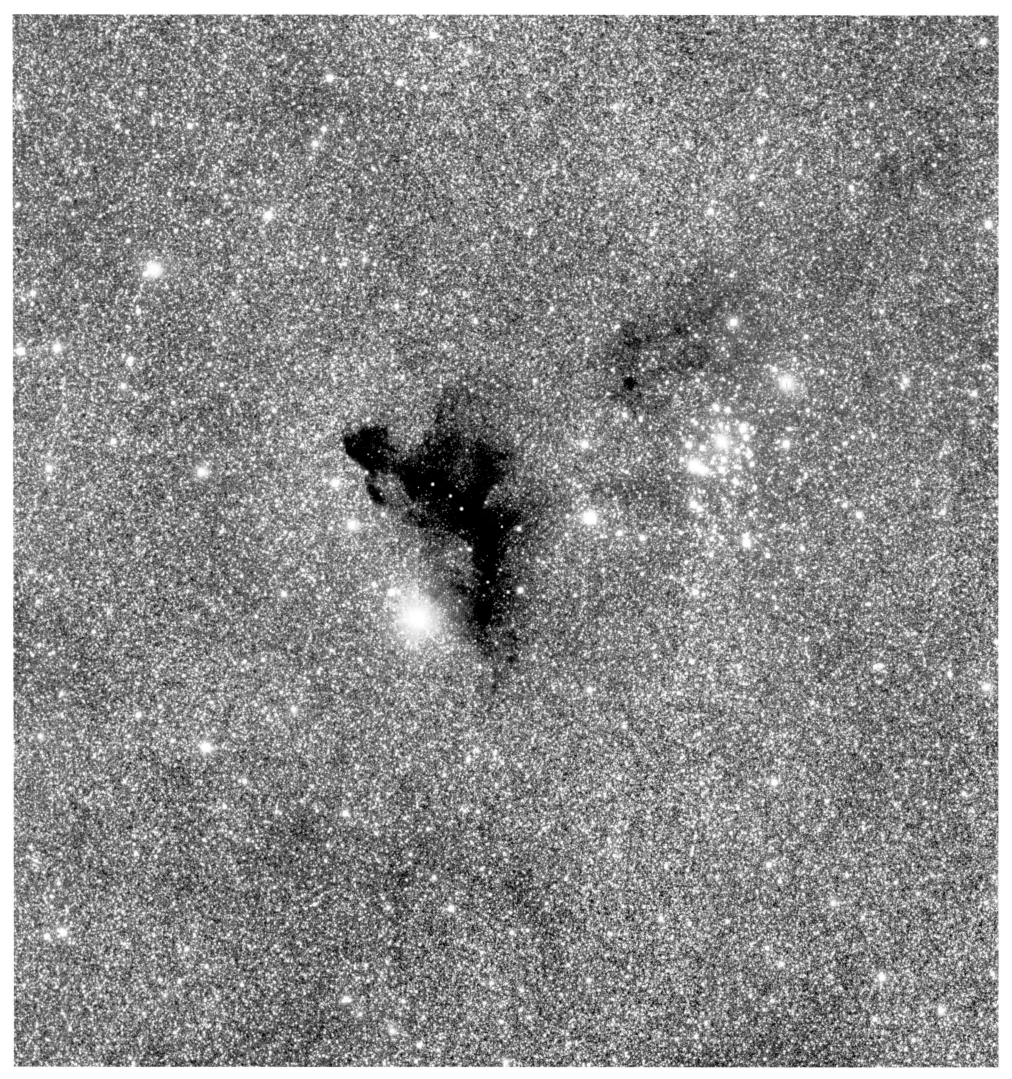

PAGES 66–67

LAGOON NEBULA

(MESSIER 8 / NGC 6523), TRIFID NEBULA AND NEBULA IC 4678

PAGE 68

LAGOON NEBULA, MESSIER 8 / NGC 6523

(KITT PEAK NATIONAL OBSERVATORY)

In Sagittarius, 6,500 light-years away, three beautiful nebulas compete to be the most beautiful. In the first photograph is a string of pink clouds, also shown on pages 70–71. At the top right shines the Trifid, a blue reflection nebula, side by side with a red emission nebula. In the second picture extends the enormous Lagoon Nebula, a site of stellar births with a diameter of 60 light-years. Its flamboyant red color comes from fluorescence of the hydrogen gas that it contains. The hydrogen is excited by the intense ultraviolet bombardment coming from the burning stars associated with the nebula. These stars are gathered into a beautiful open cluster that is perfectly visible in the close-up image, like a handful of jewels displayed on luxurious fabric. The stars are born in the very belly of the nebula—more specifically, in the small black structures that appear as shadows, which are called Bok globules.

TECHNICAL DATA

Second image taken by the 13-foot (4 m) Mayall telescope at the Kitt Peak National Observatory, Tucson, Arizona, in 1973.

PAGE 69

LAGOON NEBULA, MESSIER 8 / NGC 6523

(HUBBLE SPACE TELESCOPE)

The extremely detailed view provided by the Hubble Space Telescope has made it possible to plunge into the core of this matrix—into its most luminous part, the Sablier zone. The telescope reveals that in this region, clouds colder and dustier than the surrounding environment (gray in the image) are animated by turbulent motion. The two "pillars" (at the lower left and center of the image) are formidable cosmic tornadoes half a light-year in length. The intense stellar wind coming from the nearby blue supergiant Herschel 36 (clearly visible at the bottom right) is believed to be responsible for these towering disturbances.

TECHNICAL DATA

Color-coded image from a combination of pictures taken in July and September 1995 with the Hubble Space Telescope's Wide Field and Planetary Camera 2 (WFPC2). Three narrowband filters were used; the red light in the picture comes from sulfur ions, the blue from oxygen II ions, and the green from ionized hydrogen

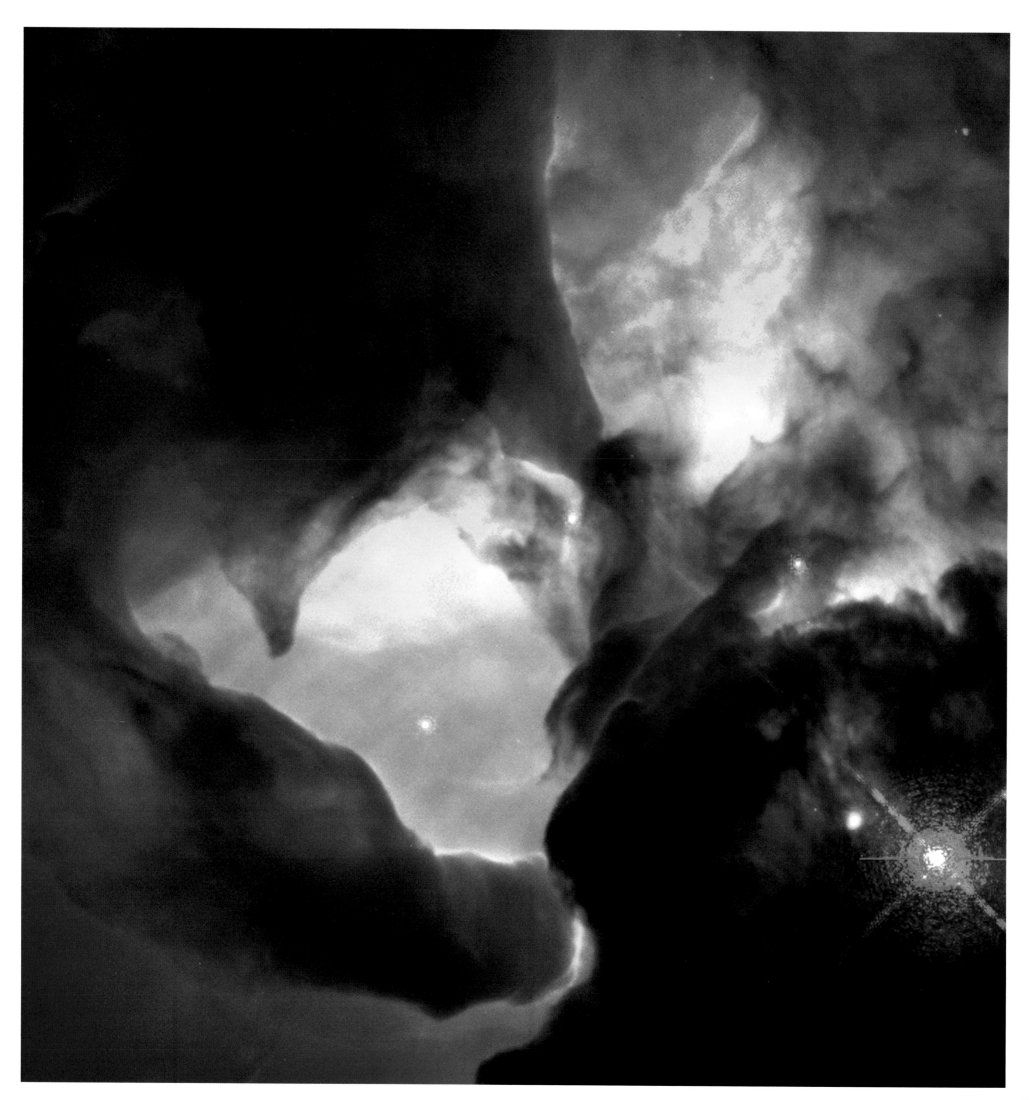

Nebula IC 4678
(CANADA-FRANCE-HAWAII TELESCOPE)

In this zone of Sagittarius the stellar density is very high. We are looking into the very heavily populated central region of our galaxy. This image is a close-up view of the chain of nebulas that appear in the upper left of the Lagoon Nebula (see picture on pages 66–67). These gaseous clouds form an enormous complex of interstellar matter. Besides ionized gases (in pink), this zone shelters numerous dust clouds. Some of them appear blue because they are reflecting light from nearby stars, while others form large spots (as at the bottom left of this view) or even filaments whose wavy silhouettes stand out against the background of stars (top right). We know today that these dust clouds can condense to form clumps within which stars develop.

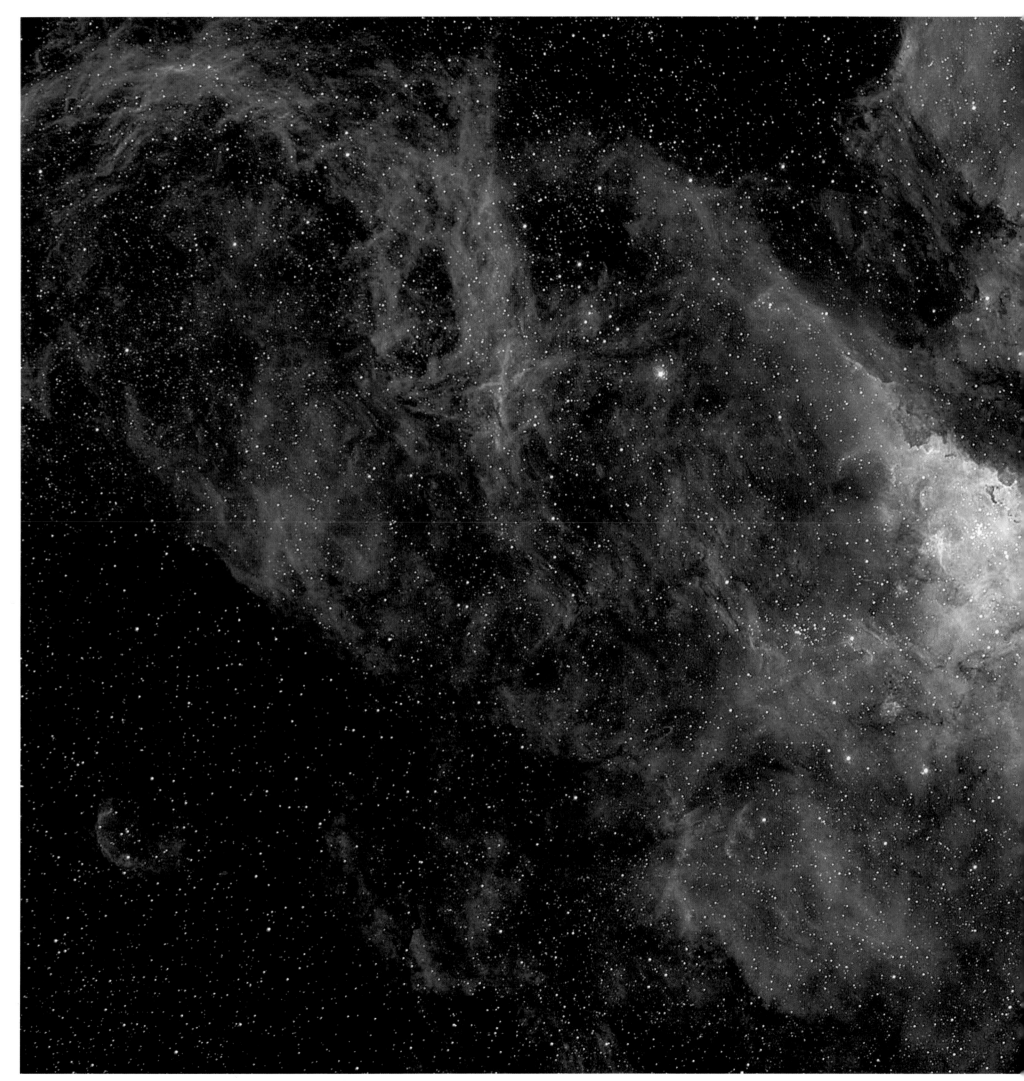

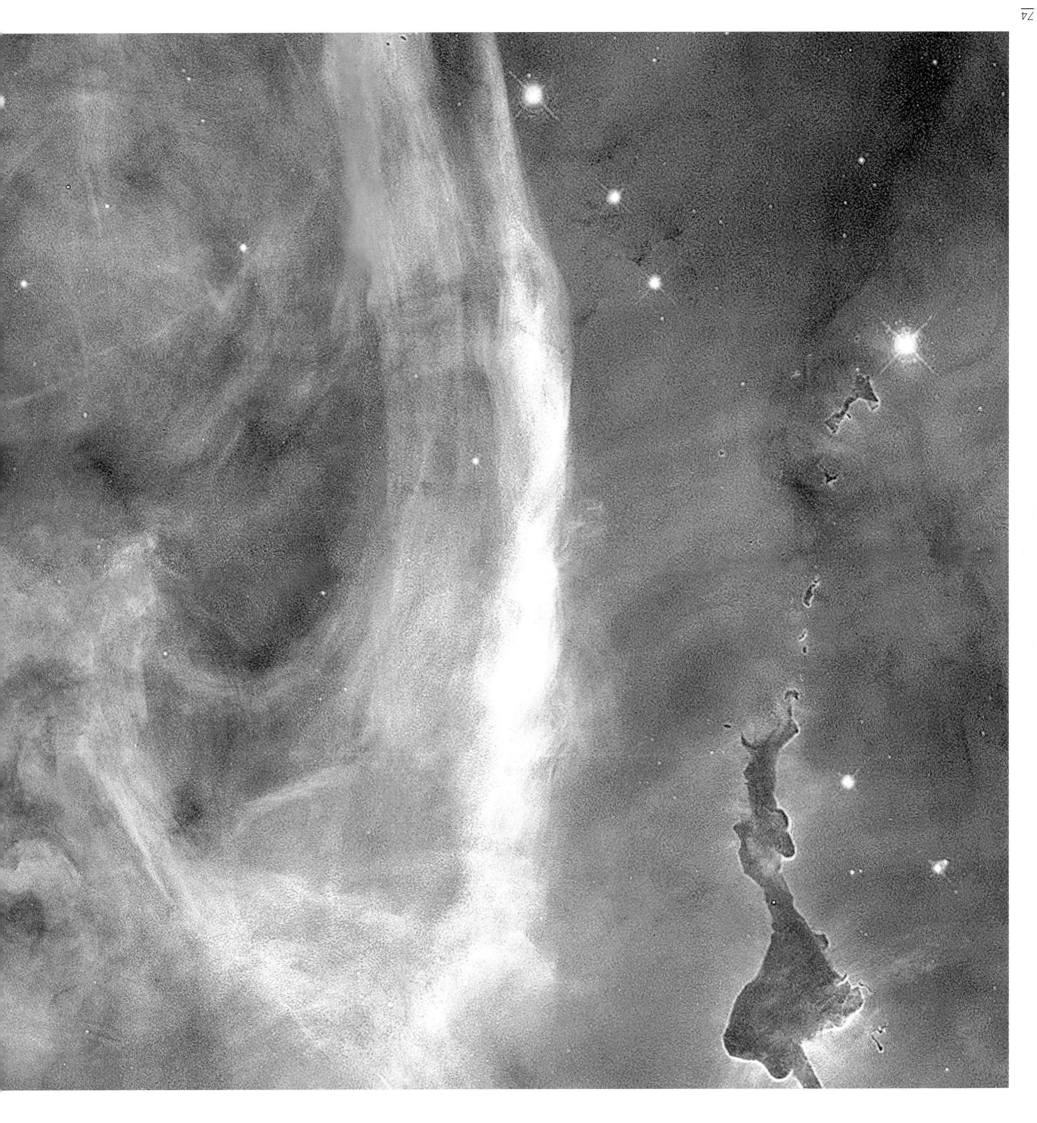

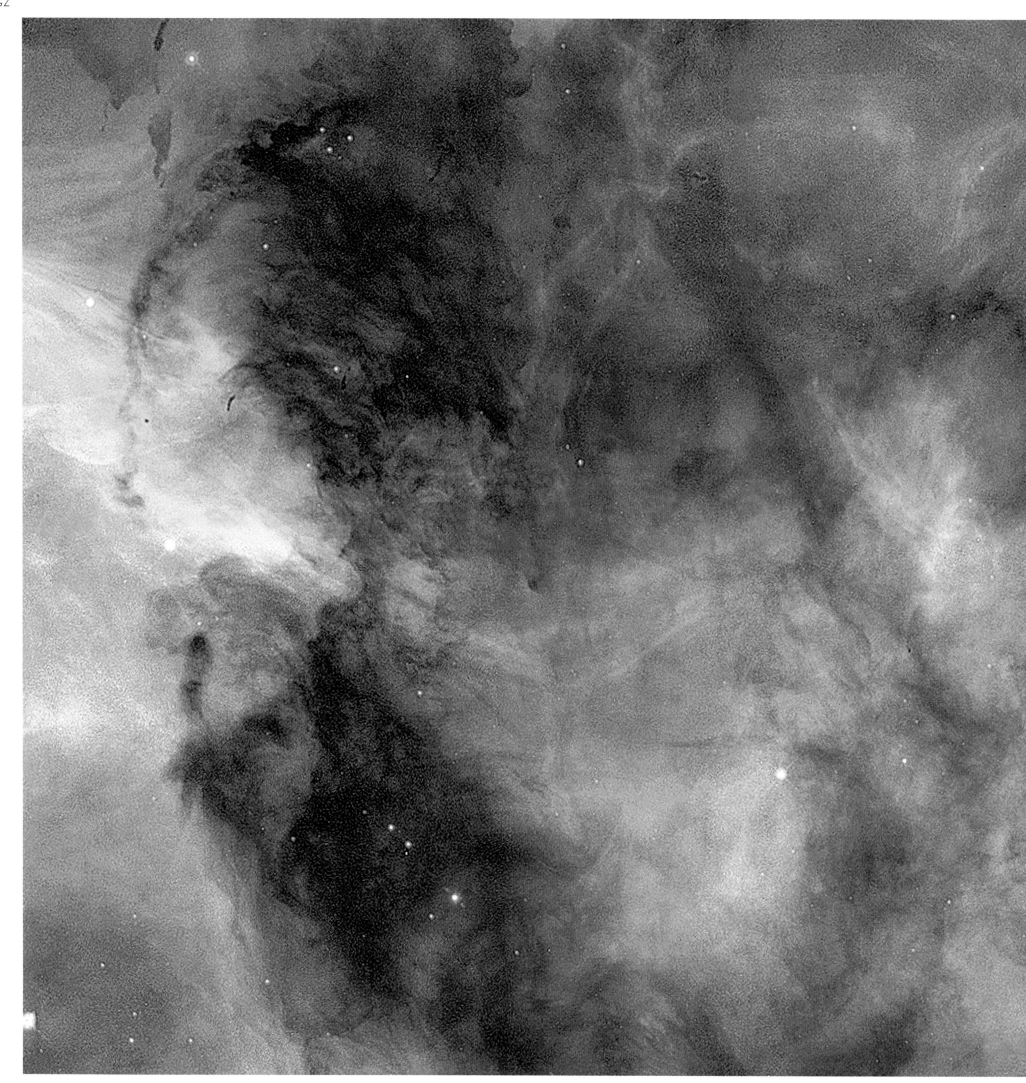

GREAT NEBULA IN CARINA, NGC 3372
(CERRO TOLOLO OBSERVATORY)

This very beautiful nebula of the southern sky, 8,000 light-years away, is one of the largest regions of star formation floating in our galaxy. Several million years ago it gave rise to a true stellar monster, the star Eta Carinae (seen in the center of the picture, to the left of the dark arc), one of the most massive and luminous ever observed, and one of the most unstable, too. For several months in 1843 it became the second or third most brilliant star in the sky, a body that blazed like a million Suns; then its luminosity dropped by a factor of one thousand. Today we suspect that the star had a nearby companion from which it had captured part of its matter. Suddenly the gas exploded violently to form the two orange-pink protuberances that obscure the star. Undoubtedly it will undergo new convulsions before it disintegrates as a supernova.

The colors in this and the following image are not realistic; they have been coded to reveal the contents and temperatures of the different gaseous layers. The blue in the picture comes from the fluorescence of oxygen, the hottest gas; the green emanates from hydrogen; and the red from sulfur, the coolest of the three.

TECHNICAL DATA
Composite of several exposures taken with the Curtis-Schmidt telescope of the Cerro Tololo Observatory in the Chilean Andes.

KEYHOLE NEBULA IN CARINA
(HUBBLE SPACE TELESCOPE)

This image is a very close-up view of the center of the Keyhole Nebula—part of the Great Nebula in the previous image— examined by the Hubble Space Telescope in the hope of revealing stellar embryos. Eta Carinae is located at the top left, outside the field of view. Several interlaced or superimposed gaseous clouds are distinguishable; the light arc located on the left is formed from very hot, turbulent gases.

Two imposing clouds that are colder and darker—rich in molecules and dust—can be made out at the center bottom and top left of the image. The first is partially immersed in the hot cloud and the second floats in front of it. Also present (top left) are some dusty globules, one in the shape of a seahorse, its head pointing toward the luminous star outside the frame. With its stellar wind the star illuminates and sculpts the shape, which may fragment to give birth to new stars.

TECHNICAL DATA
Taken April 18, 1999, with the Hubble Space Telescope's WFPC2, using six filters: F439W (blue), F502N (O III), F555W (green), F656N (H-alpha), F673N (S II) and F814W (infrared).

PAGES 78–79

CAT'S PAW NEBULA, NGC 6334
(CANADA-FRANCE-HAWAII TELESCOPE)

This complex nebulous giant floats in our galaxy 5,500 light-years away, in the direction of the constellation Scorpius. Its flamboyant, dazzling red comes from the fluorescence of gaseous hydrogen, which forms most of its substance. More than seven young, very brilliant supergiant stars, located near the cloud or tucked away inside it, trigger the cloud's fluorescence. The dark zones are curls of gas and suspended dust that are colder than the surrounding hydrogen. Astrophysicists strongly suspect that certain regions within these dark areas condense to form new stars, as in the Orion (pages 52, 54 and 55) and Eagle (pages 86 and 87) nebulas.

PAGES 80–81

COCOON NEBULA, SH 2-125

This delicate nebula, invisible to the naked eye, looks like the corolla of a flower sprinkled with pollen. It is a seat of star birth. The small group of bodies floating above it was formed about 100,000 years ago by the condensation of dark gases that it contains. These stars make up the open cluster IC 5126.

Part of the light from these very young stars is reflected from layers of dust suspended in the vacuum, diffusing a blue glow. But the majority of the energy that the stars emit causes the red-pink fluorescence of hydrogen. The largest star, in the heart of the nebula, is the source of a violent stellar wind that has blown away and hollowed out a large measure of the original gases. The process of evaporation will continue until finally this beautiful object will be disfigured and then dispersed into the interstellar medium

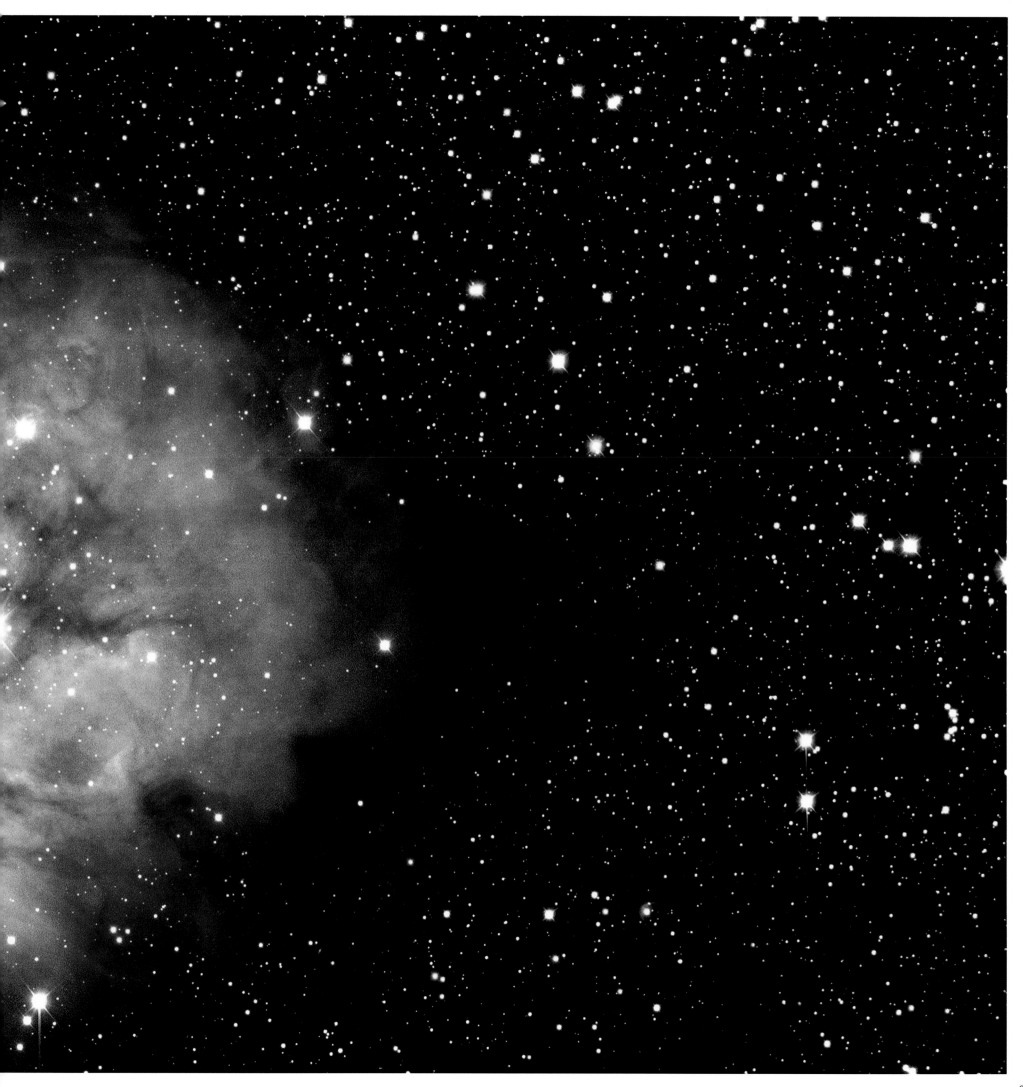

NEBULA NGC 3603, OPEN CLUSTER AND STAR SHER 25
(HUBBLE SPACE TELESCOPE)

This scene is extraordinary. It takes in the full cosmic cycle through which the stars move, from life to death and from death to life. We can actually make out stellar embryos—stars in their very early infancy—another that is dying, and also the gas clouds that accompany both the beginning and end of the life cycle.

The entire right part of the image is occupied by a stellar nursery—a cloud of cold gas composed principally of hydrogen (in yellow). Giant pillars form under the action of stellar winds (as in the Eagle Nebula, on pages 86 and 87) and release "droplets" similar to Bok globules (the black nebulosities at the top right of the image and the two spheres at the bottom center). Observers think that these structures are cocoons, concealing stars that are still forming.

To the left of this cloud is a cluster containing about 2,000 newborn giant blue stars. They ignited all at the same time, only two million years ago, in a huge explosion.

While these stars consume their youth, Sher 25, a blue supergiant star, is dying nearby (slightly to left of center). This type of star is very rare in today's universe, representing only one in 10,000. A true monster, it is 30 to 50 times larger than our little Sun, and shines 30,000 to one million times brighter! Unfortunately, this profligate use of energy can last only a short time—a few tens of millions of years, more or less (the Sun will shine a thousand times longer). Sher 25 shows other signs of its impending explosion. It has already expelled part of its own substance to form a ring of gas and two bright lobes (top right and bottom left). Their differences in color from the nebula confirm that these emanations contain heavy elements (nitrogen, carbon, oxygen, etc.) that were forged in the heart of the star. These elements will enrich the nearby clouds, which are composed mainly of hydrogen and are the source of new stars, allowing some planets to form around them.

The celestial region shown here has a size of about 30 light-years. It is located in an arm of Carina, about 20,000 light-years from the Sun, and is visible only from the Southern Hemisphere.

TECHNICAL DATA
This real-color image was recorded March 5, 1999, by the Hubble Space Telescope's WFPC2.

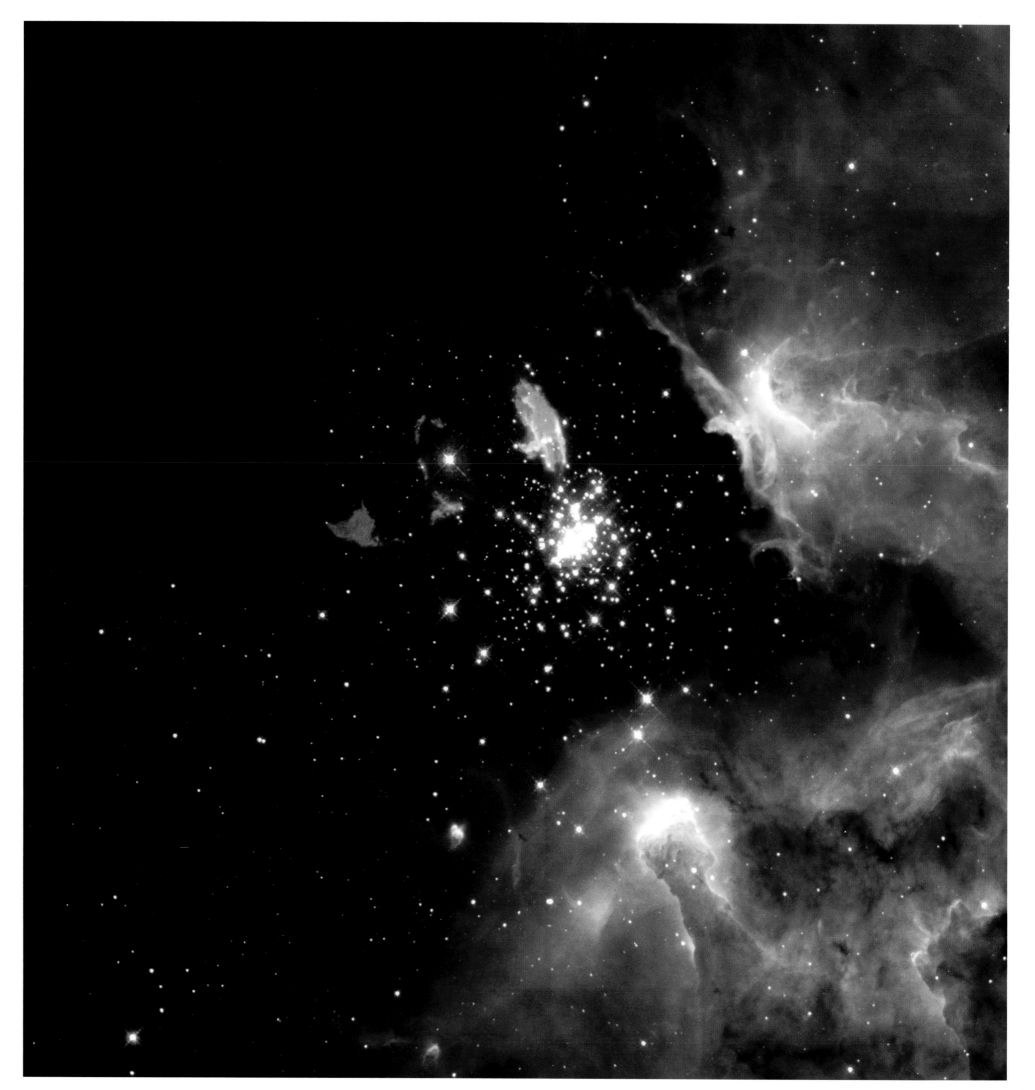

PAGE 85

North America Nebula, IC 7000, and Pelican Nebula, IC 5067/70

The North America emission nebula spans nearly a hundred light-years. In the sky it takes up an area as large as four full Moons. It is located to the left of the very brilliant star Deneb in the constellation Cygnus, a region that appears very rich in stars because it is in the line of sight of the galactic plane. On the clearest nights it can be made out with the naked eye, appearing as a pale glow. A pair of binoculars is needed to reveal the nebula's distinctive form—the "Gulf of Mexico" and the "Atlantic Ocean" are dark dust clouds. On the right a pelican with a long beak displays its left profile.

EAGLE NEBULA, MESSIER 16
OVERALL VIEW *(KITT PEAK NATIONAL OBSERVATORY)*
AND DETAIL VIEW *(HUBBLE SPACE TELESCOPE)*

The stars of a young open cluster hover, like a cloud of pink gems, over the Messier 16 nebula, the body that gave rise to them. Now they are sculpting it, and will end by tearing it to shreds.

On April 1, 1995, the Hubble Space Telescope focused on the three enormous pillars of cold gases and dust right in the center of the nebula; the previously unseen details it captured are revealed in the second picture. The longest of the protuberances reaches a light-year in length. The stars in the cluster, with their corrosive stellar wind, are causing photo-evaporation of the columns. The lighter gases escape in ethereal twists and turns, while the denser regions seem to resist breaking apart. Careful examination of the top of the left-hand column reveals tiny dark protuberances. Some of them have detached to form ovoid globules. These latter are thought to be the cocoons within which a gas can, under specific conditions, collapse in on itself and give rise to embryonic stars. They are called evaporating gaseous globules, or EGGs. Later observations showed that 11 of the 73 EGGs detected by Hubble contained many stars ready to leave the nest.

TECHNICAL DATA

Image 1: Wide-field view taken by the small 3-foot-diameter (0.9 m) telescope of the American National Science Foundation (NSF) installed on Kitt Peak, Tucson, Arizona. Three filtered images have been superimposed to obtain this picture: hydrogen alpha (green); triply ionized oxygen (blue); doubly ionized sulfur (red).

Image 2: Superimposition of three views taken with the Hubble Space Telescope's WFPC2. Singly ionized sulfur is in red; hydrogen (usually red) in green; doubly ionized oxygen in blue (usually green). Zones where hydrogen and oxygen mix appear blue-green.

LITTLE TARANTULA, OR IRIS, NEBULA, NGC 7023

The constellation Cepheus shelters this beautiful reflection nebula. At its center, HD 200775—a very young star three to fifteen times more massive than the Sun and hatched scarcely five thousand years ago—emits its first rays. They reflect from a dust cloud stretching six light-years in diameter, which would have hidden the star if the latter had been weaker or more distant.

The blue color of the nebula relates to the dust that it contains. Because they are about a micron in size, the dust particles act like microscopic filters, absorbing rays of all wavelengths except for blue, which they reflect in all directions. This optical phenomenon does not occur solely in the remote heavens, but also in Earth's atmosphere, which is laden with invisible dust and gases—that is why the sky lit by the Sun is blue

PAGE 90

Detail of Cone Nebula, NGC 2226
(HUBBLE SPACE TELESCOPE)

Astrophysicists used NASA's space telescope to obtain this very close-up view of a stellar nursery located 25,000 light-years from the Sun in the constellation Monoceros. Millions upon millions of years of gusting stellar winds and ultraviolet bombardment (from stars outside the field of view, above the cone) have resulted in erosion of gas from a vast nebula, sparing this "finger" of dust. The process is still under way, as shown by the movements of matter within the halo of fluorescent hydrogen that surrounds the object. Short-lived swirls of gas are evaporating into the vacuum, forming delicate arcs such as those in the top left of the image. This swirl could contain 65 Solar Systems.

The evaporation process will continue until the regions of highest density in the cone are exposed. They might collapse into themselves, forming gaseous spheres dense enough for nuclear reactions to start up within. Stars, perhaps accompanied by planets, could then be born.

The region shown in this image spans a width of 2.5 light-years. The three large stars visible at the top center of the image are in the foreground, not behind the cone, which is reflecting their light.

TECHNICAL DATA
Image obtained with the ACS camera on the Hubble Space Telescope, April 2, 2002. A combination of filtered images was used to isolate blue and near-infrared radiation, as well as radiation from the hydrogen alpha emission line.

PAGE 91

Ghost Head Nebula, NGC 2080
(HUBBLE SPACE TELESCOPE)

A Japanese text says, "Those who fear ghosts are those who are afraid of the night." This cosmic specter extends over an area 55 by 55 light-years in size. It is located not in our galaxy but in the one closest to ours—the Large Magellanic Cloud, in the enormous nebulosity complex of Dorado, the most active star-formation zone in the local group of galaxies.

To reveal the star-formation mechanisms of the area, observers placed filters on the space telescope's wide-field camera in order to analyze the composition of the gases and dust. The picture they obtained, shown here, reveals (in red and blue) gaseous curls containing hydrogen heated by neighboring stars. The greenish glow to the left comes from oxygen, ionized by the powerful stellar wind from a massive star located outside the field of view, to the left. The center of the ghost appears white because all the gases are present there and their light emissions are combined.

The "eyes" reveal the presence of stars concealed behind the ghost's veil. It is assumed that they were formed less than 10,000 years ago, because they are still prisoners in their gaseous womb—strictly speaking, their birth has not yet occurred.

TECHNICAL DATA
Image in enhanced colors obtained with Hubble Space Telescope's WFPC2, March 28, 2000.

PAGE 93

Elephant's Trunk Nebula, IC 1396

This cosmic pachyderm's head is a detailed view of an enormous cloud of gases and dust 3,000 light-years away, in the constellation Cepheus. Its apparent diameter is as wide as 10 full Moons, but the clouds themselves are too faint to be seen with the naked eye. The region shown here is lit by young blue and white giant stars, scarcely out of their infancy, formed within the body itself. IC 1396 is a veritable stellar nursery, like similar objects in the galaxy such as the Orion, Eagle and Cat's Paw nebulas.

This type of nebula is composed of frosty clouds of atomic hydrogen mixed with dust—the HI regions—which appear dark in these images, and hotter clouds in which the hydrogen is ionized (the energy it receives causes it to lose its only electron)—the HII regions. The latter are colored because ionized hydrogen emits red radiation when it recombines. The ionization is induced by ultraviolet rays emitted by giant or very large stars that float on the edges of the gaseous curls, or even within them. The hottest and most massive stars—called type O, or blue supergiants—shine as bright as 100,000 Suns. They ionize gas over a very large area, in the order of 350 light-years, whereas type A stars shine like 20 Suns, ionizing matter up to only two light-years around. The latter distance is less than half that separating the Sun from the closest star.

STAR CLUSTERS

Here and there in interstellar space shine brilliant swarms of stars, which are produced in great numbers. When they break clear of the cloudy cocoons that brought them forth, they remain close together for a short time, not far from their nebula of origin, tinting it with a thousand fires. But soon the young stars assert their independence and spread out into space, drifting to the four corners of their galaxy.

There is an exception to this scenario. The largest and most spectacular of these stellar swarms go through time and space in their original form, as a sphere. The stars they are made of are so close together that they attract each other with an intensity from which there is no possibility of escape. These clusters are true cosmic fossils—formed at the beginning of the universe, they contain the oldest stars known

Globular cluster Messier 80

Globular cluster Messier 80
(HUBBLE SPACE TELESCOPE)

This swarm of stars, located 28,000 light-years from Earth, is one of the most densely populated of the 147 globular clusters in our galaxy. Made up of several hundreds of thousands of stars, it formed shortly after the Big Bang, nearly 15 million years ago, when the universe looked totally different from the way it does today. Then it was primitive matter, a mixture of hydrogen and helium, probably strewn with gigantic clouds of gas in the process of collapsing in on themselves to form clusters and galaxies.

In the past this cluster was even more densely populated, and therefore more brilliant than today, but a large number of the original stars are long since dead. The red giant stars visible in this image are themselves growing old. However, researchers have recently discovered that in its center this cluster includes blue stars that are much younger and hotter. It is supposed that they came about from collisions between stars very close to the center of the cluster. In other words, the cluster was the scene of an event of great interest to astronomers—a nova outburst, a nuclear explosion on the surface of a white dwarf that had drawn matter from a nearby star into itself.

TECHNICAL DATA

High-resolution image obtained by combining two separate exposures using the Hubble Space Telescope's WFPC2 during an Italian–American study of blue clusters and Canadian–American research on dwarf novas.

PAGES 96–97

The Pleiades cluster, Messier 45
(DAVID MALIN, SCHMIDT TELESCOPE, ROYAL OBSERVATORY)

These seven magnificent stars are visible to the naked eye in the constellation Taurus. They have been celebrated since antiquity, for example, by the poet Hesiod, who lived during the 8th century BC, and by Homer, in the *Odyssey*; the books of Job and Amos in the Bible mention their existence. Their names— Alcyone, Maia, Taygete, Electra, Merope, Celaeno and Sterope— are the names of the seven daughters of Atlas and Pleione, whose namesakes are also part of the cluster. The brightest star, Alcyone, shines like a thousand Suns.

The stars sail along 450 light-years from Earth. With binoculars, not only are the seven stars visible, but dozens of others emerge from the darkness of a celestial region twice as broad as the full Moon. Galileo saw 36 with his telescope in 1610. Since then it has been discovered that this open cluster contains over 3,000 young blue stars, all born 70 million years ago. For a long time the thin veil that surrounds the stars and reflects their light was thought to be the remains of the dense nebula that gave birth to them. Today measurements suggest that this diffuse nebulosity actually belongs to another cloud that happens to be in the line of sight.

This cluster will dissipate over the next few million years, and each star will fly off to follow its own path in the galaxy.

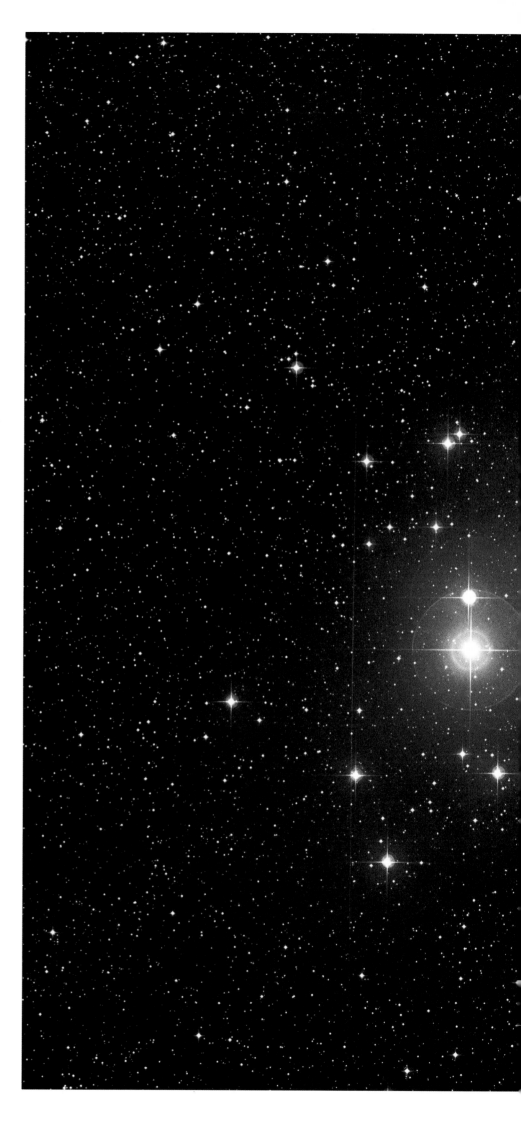

Open stellar clusters Messier 35 and NGC 2158

(CANADA-FRANCE-HAWAII TELESCOPE)

Located 2,800 light-years from Earth, this swarm of blue stars seems to be launching itself into space as if to rejoin the second stellar grouping, denser but paler, at the top left of the image. Their chance positions in the image place these two clusters near each other, but in fact they are very far apart and, furthermore, very different in age and content. The vivid blue stars form the younger cluster M35. They can be distinguished through binoculars, or even with the naked eye, on a particularly clear winter night, appearing as a pale spot at the foot of the hexagon of Gemini. The cluster contains 2,500 stars grouped in a sphere 30 light-years in diameter. These bodies have only recently emerged from the gaseous cocoon that gave birth to them 150 million years ago, a vast cloud of gas and dust that has since disappeared.

The second cluster, at the top left of the image, is six times more distant than M35 and ten times older. At the beginning, many blue giant stars populated this cluster as well, but they have since disappeared—such stars have a short life expectancy. Very massive and extremely bright, they exhaust their nuclear fuel in a few tens of millions of years. Two types of brilliant stars today dominate this cluster. Old yellow stars, similar to our Sun, are less luminous than the blue giants, but their life expectancy reaches 10 billion years or more. The second type is dying stars, the red giants easily visible in this image.

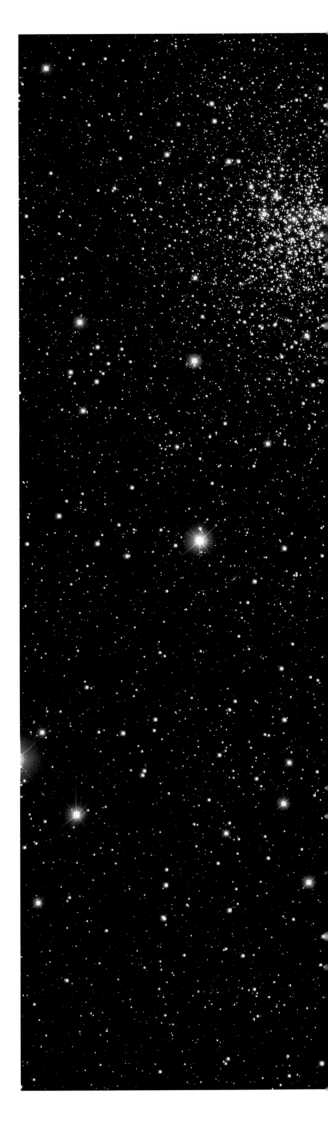

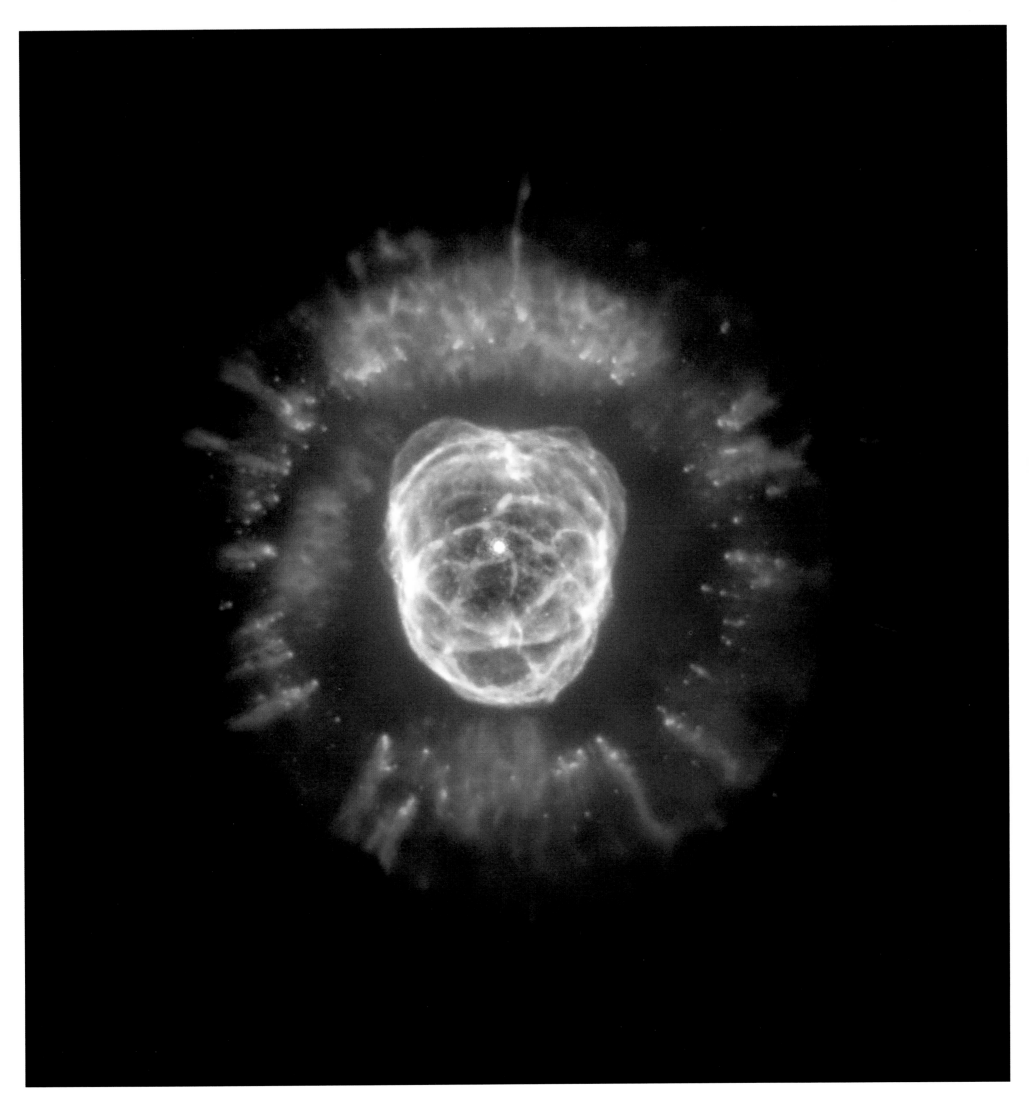

PLANETARY NEBULAS: THE END OF MEDIUM-SIZED STARS

All that lives must die. After having shone without fail for 10 billion years, stars like our Sun end their lives as veritable cosmic fireworks. When they have consumed all their original nuclear fuel of hydrogen and helium and forged atoms of carbon, nitrogen and oxygen, their fire weakens and the forces that kept them intact begin to fluctuate. The spent star swells beyond recognition. If there are planets nearby, they pay the price as their atmospheres and oceans evaporate into space. Then the star's gases are expelled into the space around it in convulsive bursts. As the death of the central star continues, it spreads a cold but intense light that makes the expelled shell of gases glow with a thousand fires. After 30,000 years, the heart of the star stops shining completely, and its rainbow-colored nebula melts into the darkness.

Eskimo Planetary Nebula, NGC 2392

ESKIMO PLANETARY NEBULA, NGC 2392
(HUBBLE SPACE TELESCOPE)

This striking image was obtained in January 2000 by the Hubble Space Telescope. The Shuttle astronauts had just repaired the instrument's failing gyroscopes, and to verify that the telescope was functioning correctly, NASA observers chose as their target one of the most photographed nebulas in the sky. This nebula is 5,000 light-years away. Discovered in 1789 by William Herschel, it can be observed with a small telescope during the early evenings of winter, high in the skies of Gemini. It resembles a human head hooded by a parka, or perhaps a clown's head with a ruff. Like all the planetary nebulas, NGC 2392 is made up of matter expelled into space by the remains of a dying star.

But this nebula is particularly complex and unusual. The central star (the "nose") is unusually bright. The outer envelope (the "hood") would have formed about 10,000 years ago, during the star's red giant phase. Its diameter has since reached one light-year. The "fur" is orange filaments, probably being swept toward interstellar space by the intense stellar wind from the central star. Inside, matter expelled more recently forms spherical "clouds" that overlap in a complex structure.

Today more than 1,500 planetary nebulas have been cataloged within our galaxy, but it is estimated that it contains 10,000 in total. Such a body is the final message to us from a yellow star whose mass falls between 0.8 and 8 solar masses, and whose life expectancy is tens of billions of years before it goes out forever. When the star's fuel starts to become exhausted, it expands by a factor of a hundred, becoming a red giant with a diameter of 186 million miles (300 million km). Its shell is so tenuous that it evaporates into interstellar space, in outbursts because of the instability of thermonuclear reactions. Such a star sheds several trillion tons of matter each year. As it loses mass, it contracts and warms. The star then emits intense ultraviolet radiation, which hollows out and sculpts the previously expelled gas around it, forming streaks, globules and jets. What's more, the UV radiation causes illumination of the nebula by inducing fluorescence in its atoms—oxygen glows green, nitrogen blue or red-pink (according to its state of ionization), helium yellow or blue, and so on.

Gradually the star's gases diffuse into space and, after 10,000 to 30,000 years, the nebula ends by melting into darkness. The central star, reduced to the size of Earth, cools and grows fainter until it becomes invisible.

TECHNICAL DATA

Image captured January 10 and 11, 2000, with the WFPC2 of the Hubble Space Telescope. Colors come from the fluorescence of nitrogen (red), hydrogen (green), oxygen (blue) and helium (violet).

SPIROGRAPH PLANETARY NEBULA, IC 418
(HUBBLE SPACE TELESCOPE)

This celestial gem, named from its resemblance to figures created by the Spirograph drawing toy, marks the last act of a star in the constellation Lupus. Its envelope has been dissipated into space by successive spasms caused by the thermal pulsations to which the star is subject. This action has formed a spectacular ellipsoid in rainbow shades with a diameter that reaches some 10 light-years. To obtain this image, as for most other images of planetary nebulas, astronomers placed colored filters in front of the telescope camera in order to capture separately the fluorescence of different chemical elements present in the object. Each color indicates the presence of a specific ion. The red zones are produced by ionized nitrogen, the coldest gas in the nebula and farthest from the central star. The green areas indicate the presence of hydrogen and the blue zones signal oxygen, which can also be detected very close to the center.

TECHNICAL DATA

Composite view obtained with the Hubble Space Telescope's WFPC2 in February and September 1999, using three colored filters.

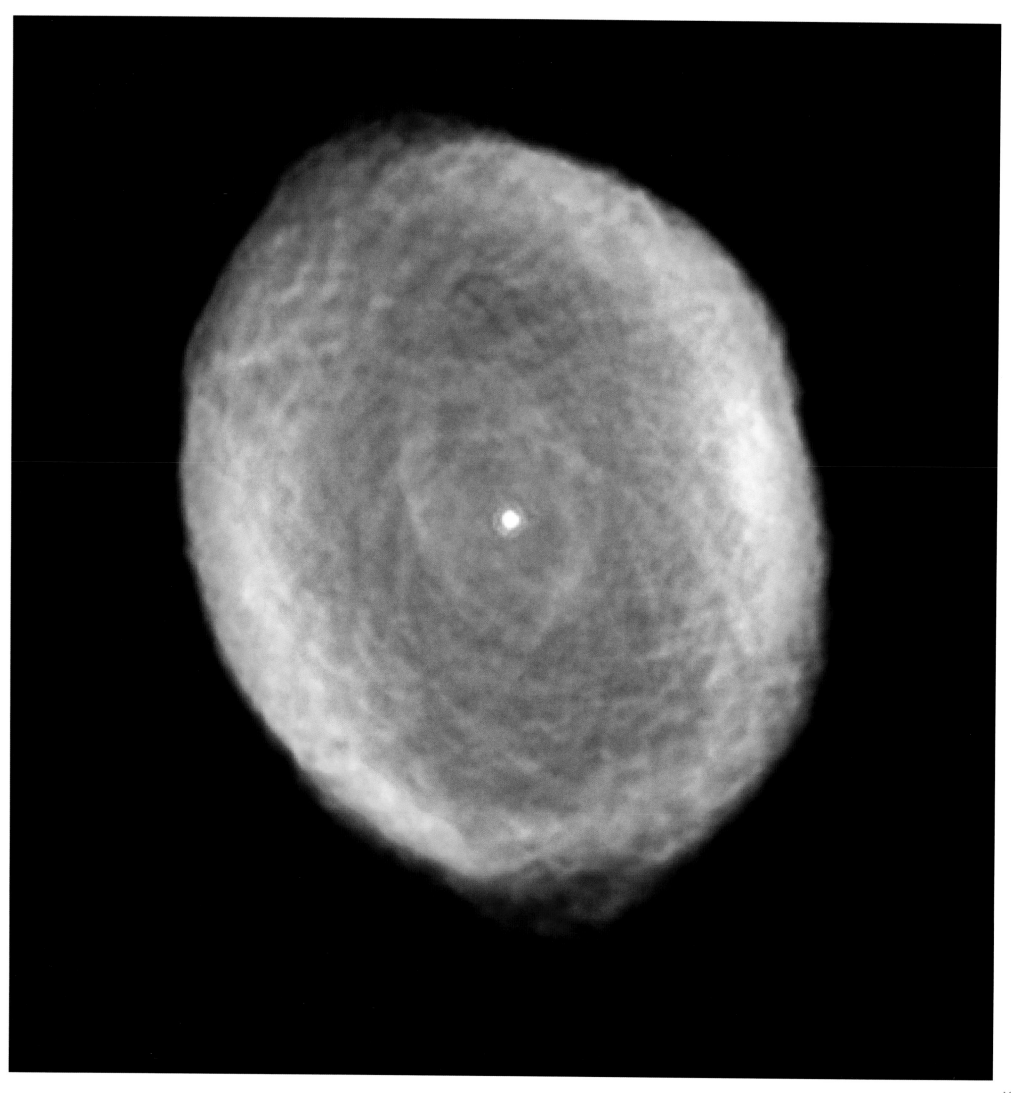

PLANETARY NEBULA ABELL 39
(KITT PEAK NATIONAL OBSERVATORY)

In 1966 this "soap bubble" was discovered in the constellation Hercules (7,000 light-years from the Sun) hiding the galaxies in the background. Slightly off-center in the nebula, the corpse of the star that gave rise to it rests in peace. It is a white dwarf, a small, still incandescent body that will slowly fade away until it becomes invisible.

How did this bubble form? As it was dying, a star the size of the Sun grew immensely to become a red giant. When the star had exhausted the nuclear fuel that enabled it to shine, it expelled its gaseous envelope into space. As it expanded, these gases formed a magnificent bubble—a perfect sphere whose diameter today has reached five light-years, a distance greater than that separating the Sun from the closest star, Proxima Centauri. However, this bubble will grow paler little by little until it melts into the cosmic darkness. (An analysis of this image has shown that the star that gave rise to the nebula contains half as much oxygen as our Sun; this fact has not yet been explained.)

TECHNICAL DATA

Picture taken for research purposes in 1997 at the focus of the 11.5-foot-diameter (3.5 m) WIYN telescope at the Kitt Peak National Observatory in Tucson, Arizona. A blue-green filter was used to isolate the light rays emitted solely by oxygen atoms (500.7 nm)

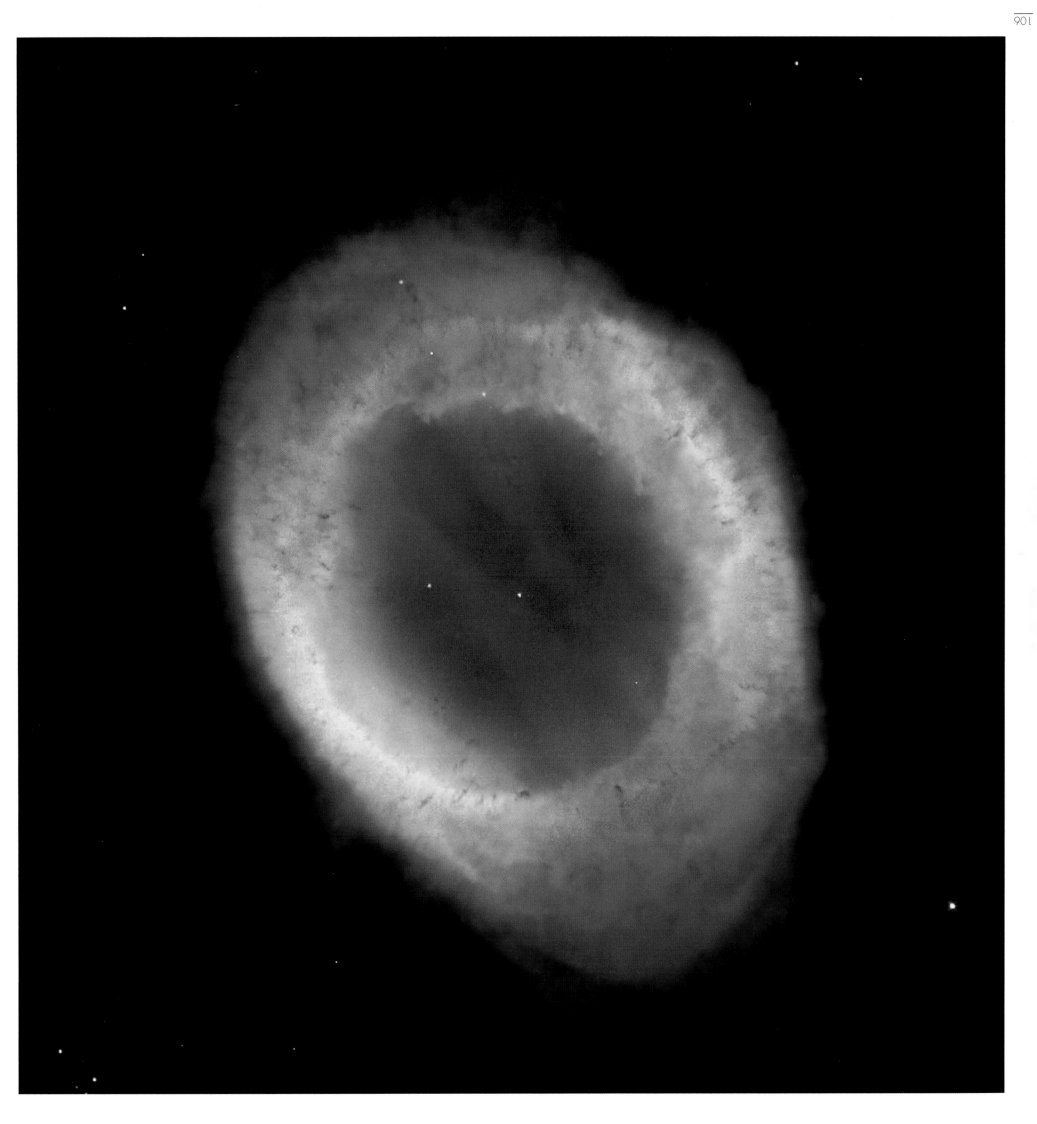

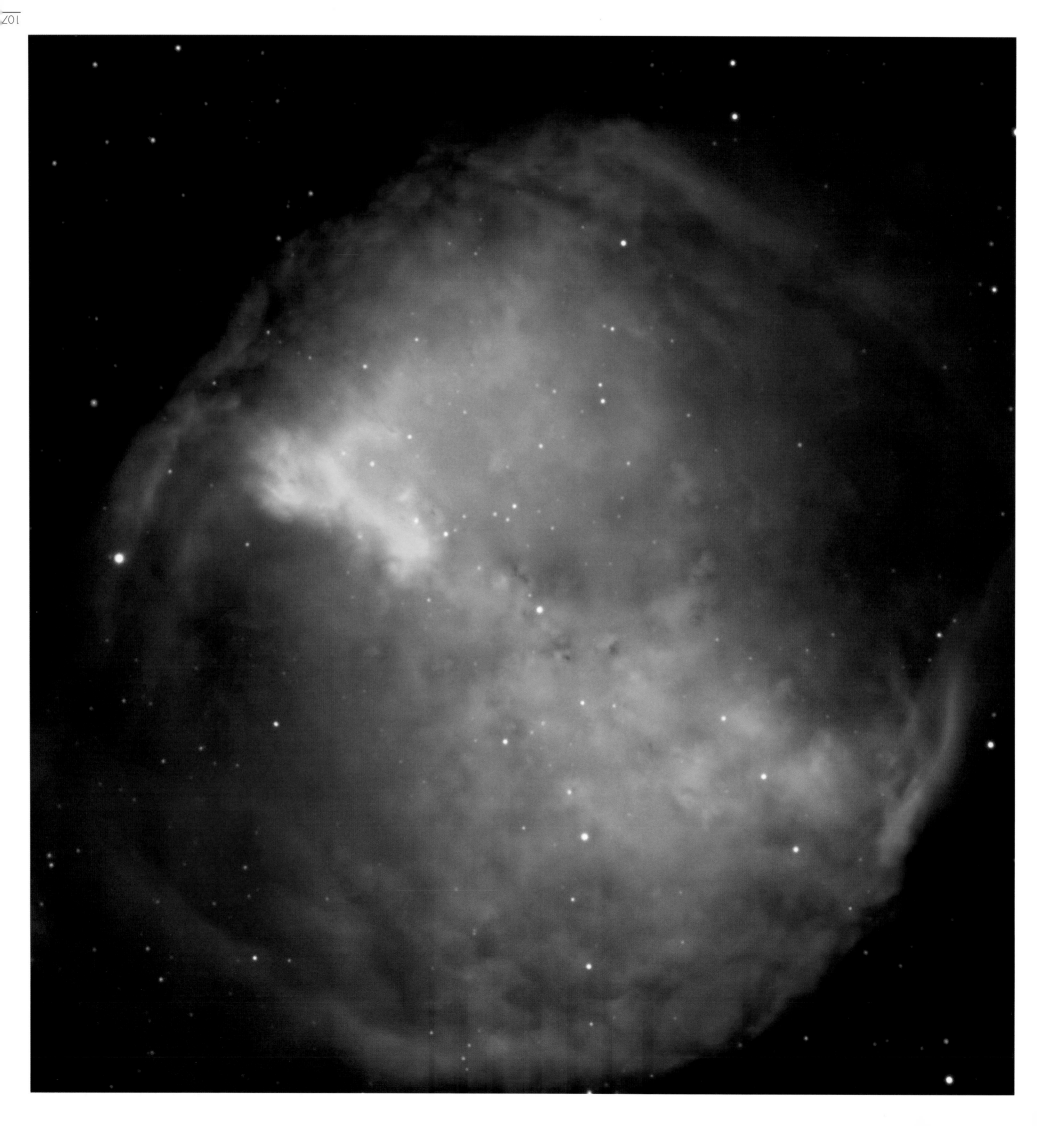

PAGE 106

PLANETARY NEBULA MESSIER 57 IN THE CONSTELLATION LYRA
(HUBBLE SPACE TELESCOPE)

This planetary nebula in the constellation Lyra is one of the most photographed in the sky. It was discovered by Antoine Darquier of Toulouse, France, in 1779, who mistook its real nature. "As large as Jupiter, it resembles a pale planet," he wrote. William Herschel, who had observed other, similar objects, classified them as planetary nebulas in 1784 or 1785, because of their resemblance to the planet Uranus, which he had just discovered. However, several years later he detected a very bright point in the center of one of them. That convinced him that he was dealing with an unknown type of body—a star associated with a light and diffuse substance. He had concluded correctly. The star had expelled some of the gases that it had itself produced. Here oxygen glows green, nitrogen red and helium blue.

Appearances, however, can be deceiving. It seems that M57 is neither an ellipsoid, as had been assumed for a long time, nor a torus (donut shape). Its real shape is more like a cylinder or an hourglass; the view we see is through the base, along its axis.

TECHNICAL DATA
Image captured by the WFPC2 of the Hubble Space Telescope in October 1998. The coding approximately represents actual colors. Filters used: red F658N (N II), green F501N (O III), blue F469N (He II).

PAGE 107

DUMBBELL PLANETARY NEBULA, MESSIER 27 / NGC 6853
(VERY LARGE TELESCOPE)

Here is the first planetary nebula ever discovered. It was described on July 12, 1764, by the French astronomer Charles Messier, while he was examining the Little Fox constellation and looking for comets. The intrinsic brightness of this body is that of a hundred Suns. Its apparent diameter is nearly that of the full Moon, so it can be observed with only a small instrument or even binoculars. Its light takes about 1,200 years to reach us.

This planetary nebula formed at the moment of a star's death about 3,000 or 4,000 years ago. What remains of it—a still very bright white dwarf—is clearly visible in the center of the image, and it is responsible for the beautifully colored finery of the expanding gas. Colored filters were used to isolate this light and reconstruct an image as aesthetically pleasing as it is valuable to scientists. This image was captured by the Very Large Telescope, the most powerful observatory in the world, which was in the test phase at the time the picture was taken. To obtain the picture, astronomers installed a wideband filter that allows blue light to pass through to the telescope's high-definition camera. Having obtained a first, short-duration exposure that way, they next proceeded to take two more separate images. The first used a filter that isolated the light emitted by doubly ionized oxygen (green) and the second used a filter specifically for hydrogen (red). The three views were then superimposed to obtain this almost realistic view of the nebula. The two colored lobes are the origin of its names Dumbbell nebula and Diabolo nebula, the latter from a two-headed top.

TECHNICAL DATA
Composite image in three colors: interference filters in the bands of oxygen and hydrogen alpha, and wideband Bessel B. Obtained September 28, 1998, under mediocre visibility conditions using the FORS1 CCD camera (2048 x 2048 pixels) in the first dome of the Very Large Telescope.

PAGE 109

HELIX PLANETARY NEBULA, NGC 7293
(CANADA-FRANCE-HAWAII TELESCOPE)

Here is the closest of the planetary nebulas, photographed from the summit of Mauna Kea, Hawaii. It is 450 light-years from Earth, a hundred times farther than the star closest to the Sun. Its apparent diameter extends a quarter of a degree (half that of the full Moon), but its actual diameter exceeds five thousand times that of the Solar System. At the center shines a nearly moribund star, a white dwarf.

The magnificent colors adorning this nebula come from nitrogen and hydrogen (red) and from oxygen (green). These gases, which originated in the outer layers of the star, are valuable because they enrich interstellar space with heavy elements such as carbon and oxygen, in the form of molecules and dust

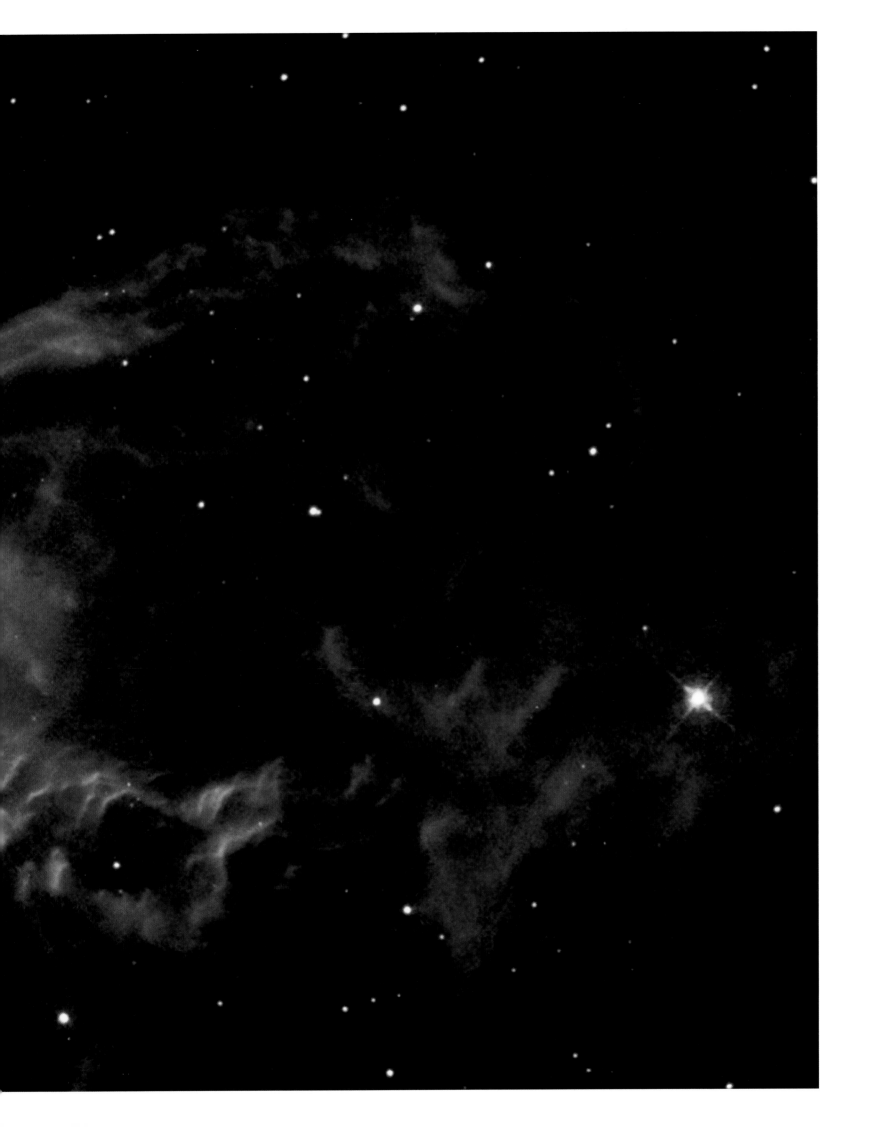

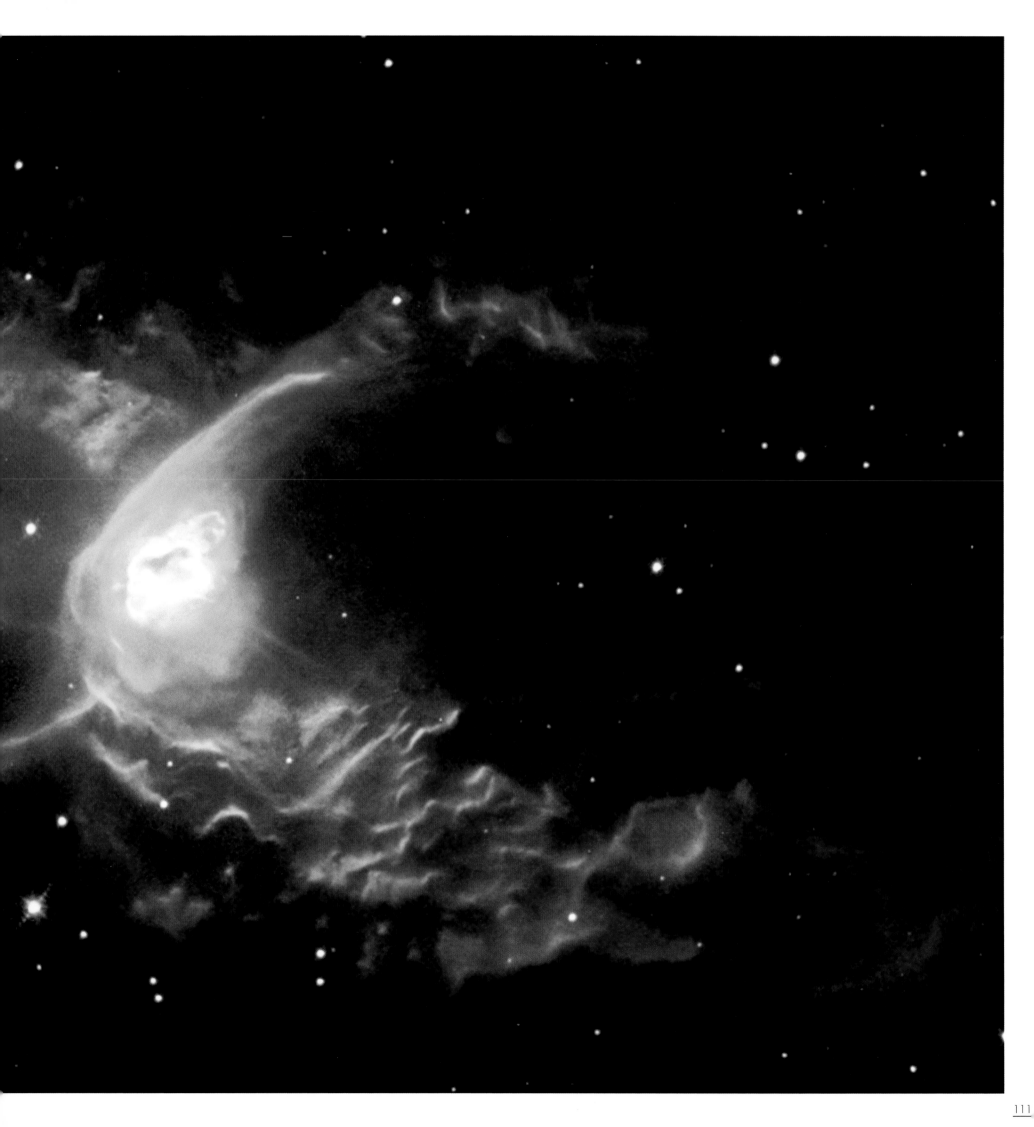

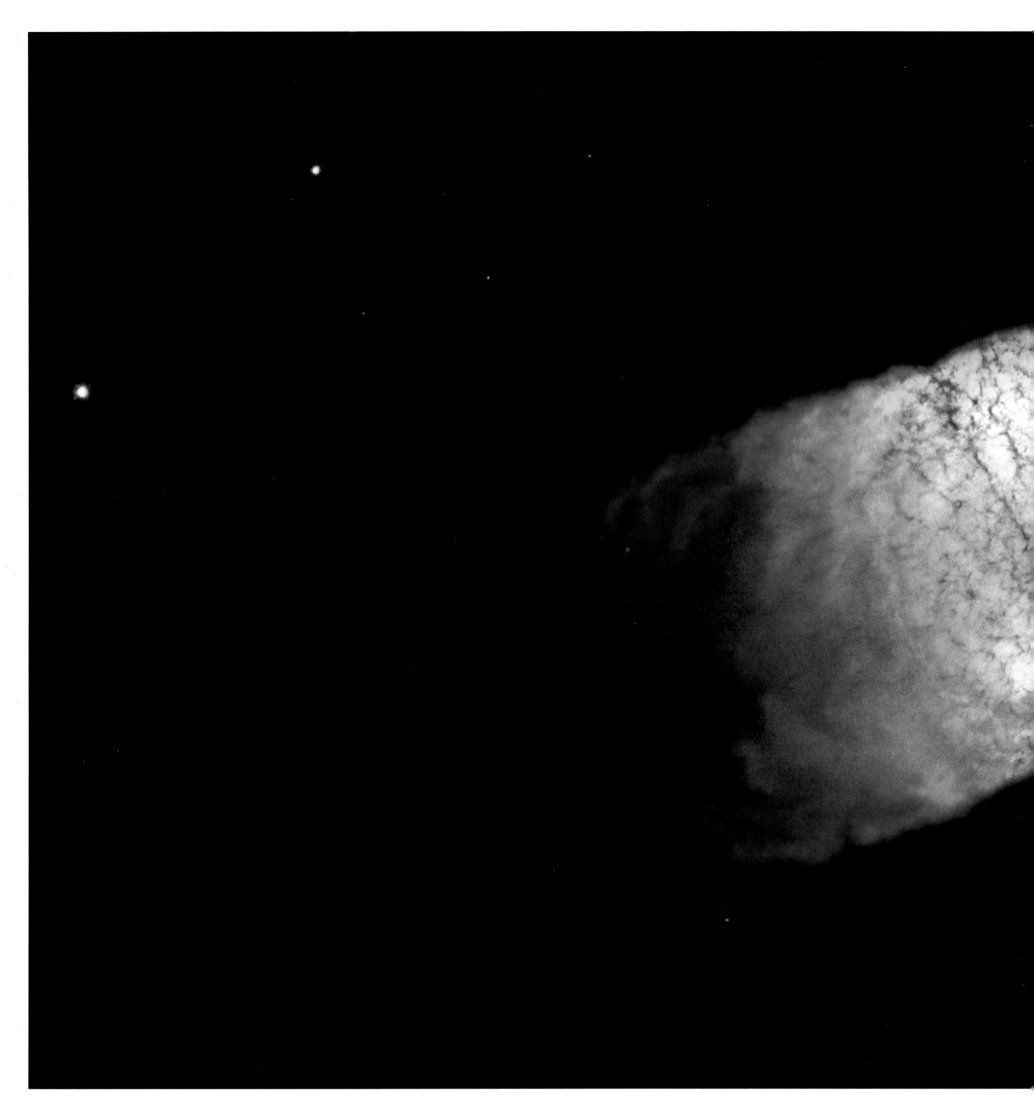

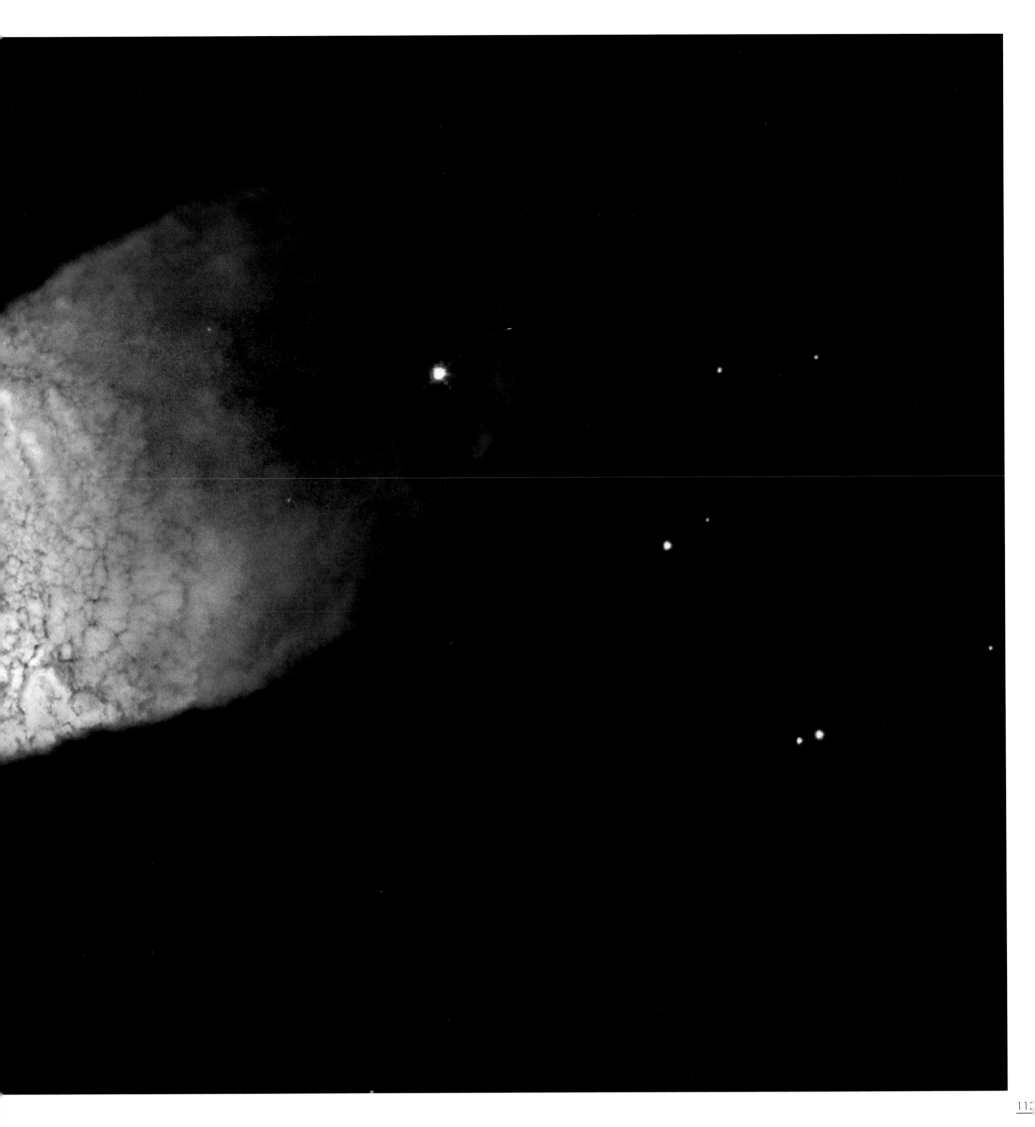

Red Spider planetary nebula, NGC 6537
(HUBBLE SPACE TELESCOPE)

At 3,000 light-years from the Solar System, in the constellation Sagittarius, the Hubble Space Telescope discovered this curious planetary nebula, called the Red Spider because of its shape. Within it lurks a particularly hot white dwarf, spewing out a flood of particles of exceptional intensity and speed—4 to 10 million miles per hour (7 to 16 million kph). This stellar wind, a hundred million times more powerful than that of the Sun, acts like a furious squall on the ocean. It raises cosmic-sized waves in the interstellar gas, hundreds of billions of miles high and spreading at a speed of 600,000 miles per hour (1 million kph). The tumult is responsible for the complex appearance of the nebula's outer arms, which are sculpted in strange waves. This phenomenon has probably been at work for a long time, as wavy fragments have ended up detaching themselves.

Another property of this nebula draws the observer's attention—the S-shaped symmetry of the opposing lobes. To explain it, astronomers cite the possible presence of a stellar companion orbiting the white dwarf. As of now, none has been detected. New observations will be necessary to determine the exact nature of the exotic bodies that lurk in the heart of this celestial arachnid.

TECHNICAL DATA

Image obtained using the WFPC2 of the Hubble Space Telescope. Five filters were used; light emitted by sulfur appears in red, nitrogen in orange, ionized hydrogen (alpha band) in green, atomic hydrogen in light blue, and ionized oxygen in dark blue.

PAGES 112–113

Retina planetary nebula, IC 4406
(HUBBLE SPACE TELESCOPE)

What is the real shape of this planetary nebula, one light-year long, located 1,900 light-years from us in the constellation Lupus? Like many other bodies in this family, IC 4406 has nearly perfect bipolar symmetry—its two parts, left and right, are mirror images of each other. From Earth we see the nebula from the side, but space travelers flying around it would be able to contemplate its face. They would then see a multicolored ring similar to that of the Lyra or Helix planetary nebula (see pages 106 and 109).

This picture reveals a turbulent network of darker gas enveloping the central zone of the nebula; each of these bruise-like marks extends for about 160 astronomical units. The density

within them is a thousand times greater than in the rest of the nebula. Their nature remains mysterious, along with their destiny. Will this network of filaments grow in volume and follow the expansion of the nebula, or will it dissipate into space?

TECHNICAL DATA

The image is a composite of data acquired using the WFPC2 of the Hubble Space Telescope in June 2001 and again in January 2002. Three filters were used to reconstruct a color image; oxygen is in blue, hydrogen in green, and nitrogen in red (all are ionized).

PAGE 115

Gomez's Hamburger proto-planetary nebula
(HUBBLE SPACE TELESCOPE)

This strange pale glow is associated with a star similar to the Sun that is running out of nuclear fuel. Here we are watching the first stages of formation of a planetary nebula as the body starts to lose its gaseous envelope and send it off into space. It shines because it contains microscopic dust particles that reflect light from the central star, which, it should be added, is not visible. Hiding it is a large band of opaque dust—probably a ring seen in profile from Earth—whose presence has not been explained. In less than a thousand years the extremely hot heart of the star (more than 10,000°C/18,000°F) will be completely uncovered and exposed to the cosmic vacuum. It will then produce an intense stellar wind that will sweep the dust far away. But this nebula will not lose its brightness quite so easily—just the opposite. The powerful ultraviolet radiation emitted by the star will provoke fluorescence of the gases around it, and the entire nebulosity will be adorned in magnificent colors.

Gomez's Hamburger is located in the constellation Sagittarius. Its light takes 6,500 years to reach us.

TECHNICAL DATA

Image acquired using the WFPC2 of the Hubble Space Telescope, February 22, 2002. The 32-minute exposure used three filters: F675W, F555W, F450W.

PAGE 117

Henize 3-401 planetary nebula
(HUBBLE SPACE TELESCOPE)

Contrary to what astronomers supposed for more than a century, planetary nebulas are not mostly spherical or ellipsoidal. The Hubble Space Telescope and the most powerful terrestrial instruments, such as the European Very Large Telescope, have revealed the existence of a veritable bestiary of shapes. Among the most surprising are the bipolar or bilobal nebulas, including this one 10,000 light-years away—one of the most distant ever observed. How could a perfectly spherical star produce two symmetrical jets of matter? Astrophysicists are obliged to imagine complex new scenarios in order to come up with an explanation.

Some specialists suspect that in the center of the nebulosity is not just one star but a couple of very close bodies. The denser one would pull shreds of stellar material from the other so that a gaseous ring—with a diameter 10 times larger than the orbit of Pluto—would form a belt around this so-called symbiotic binary star. Like a jet engine or a rifle barrel, the ring would channel the flow of gases, forcing it to escape above and below it. Other researchers suppose, however, that only the magnetic field of the dying star is responsible for the emission of these two jets. New observations are under way to shed light on this phenomenon.

TECHNICAL DATA

The image was recomposed using three exposures obtained on June 12, 1997, with the Hubble Space Telescope's WFPC2. Exposures were taken through an orange wideband filter (blue details), a filter selecting emissions of hydrogen alpha (in red), and a final filter selecting emissions of singly ionized sulfur (in green).

"ANT" MENZEL 3 BIPOLAR PLANETARY NEBULA
(HUBBLE SPACE TELESCOPE)

Here is a very strange manner of dying. In the center of this image, the star is ejecting its gases in amazing symmetrical globules and multiple jets, forming the body of an immense multicolored insect in the constellation Norma, 3,000 light-years away. These bizarre structures are the result of interactions between the stellar wind and the layers of surrounding gas, which are more or less resistant to the passage of the shockwave.

The bilobal nature of this body is similar to that of the Henize 3-401 nebula (see page 117). However, although several other nebulas with jets are known, no other planetary nebula displays bubbles of ejected matter that appear to form the body of an ant.

Four filters were used to take this picture in order to analyze the distribution of gases in the nebula. Sulfur shows its presence in red, nitrogen in green, hydrogen in blue and oxygen in blue-violet.

TECHNICAL DATA

The image results from a combination of two observations by the Hubble Space Telescope's WFPC2, the first in July 1997 and the second approximately one year later

THE DECLINE AND DEATH OF GIANT STARS

The cosmos produces its own monsters: blue stars, each one able to hold a thousand of our Suns and to shine a hundred thousand times more intensely. From the time these enormous incandescent spheres are born, their destiny is sealed—no exit, no exceptions allowed. Though they are dazzling on the cosmic scale, their life will be a thousand times briefer than the Sun's—such is the price they pay for their extraordinary brightness. Their profligate use of energy rapidly drains the enormous reserves of nuclear fuel with which they were initially endowed.

The end of such a star is cataclysmic. As its heart begins to palpitate irregularly, the supergiant star undergoes sudden episodes of expansion, causing it to lose its matter in shreds. Once its heart is exposed, the star becomes a supernova, exploding in a veritable swan song, 10 billion times brighter than its peers.

We owe our life to these stars. The oxygen that we breathe and the heavy atoms in our own cells have been forged there—some within the fiercely burning heart of the stars, others within their matter, torn apart by the final explosion. This fertile debris, strewn far and wide in interstellar space, enriches gas clouds with its atoms. The fragments of one of these stars gave rise to the Solar System.

And so, on Earth as it is in Heaven, dust returns to dust, and life comes from death.

Crescent Nebula, NGC 6888

CRESCENT NEBULA, NGC 6888
(CANADA-FRANCE-HAWAII TELESCOPE)

This nebulosity, located 4,700 light-years from Earth in the constellation Cygnus, is significant in its size—16 light-years wide by 25 light-years long. At its center shines the star WR (Wolf-Rayet) 136, a type so rare that only two hundred examples exist in our galaxy. It is a true stellar mammoth, equivalent to between 20 and 100 solar masses.

For about 250,000 years this body has been at the red supergiant stage, undergoing a harrowing death-struggle during that time. The evaporating hydrogen that once enrobed the star has partially uncovered the core, which is rich in helium. A Dantesque hell heated to several million degrees launches furious stellar winds into space. These winds reach an extreme speed of 3.8 million miles per hour (6.1 million kph), robbing the star of matter equivalent to the mass of the Sun every 10,000 years. Ever since these winds reached the previously ejected hydrogen shell, they have been subjecting it to harsh torture. The shock is causing the gaseous matter to be torn to shreds and heated, illuminating it in red through the phenomenon of fluorescence.

Soon (on the cosmic scale), WR 136 will have burned up all the helium that it contains and will no longer be able to carry on its thermonuclear processes. It will then explode as a supernova, reducing this delicate red nebula to shreds.

REMNANTS OF SUPERNOVA 1987A
(HUBBLE SPACE TELESCOPE)

At the center of this image, a strange double ring floats in the vacuum 170,000 light-years from Earth. It seems suspended above the red gas of the nebula. To photograph the scene, the Hubble Space Telescope was pointed at an area of the sky that has captivated astronomers since 1987—the Tarantula Nebula in the Large Magellanic Cloud, an irregular galaxy visible from the Southern Hemisphere. In that year an extremely rare event happened.

It all started on the night of February 24–25 at Las Campanas Observatory in the Chilean Andes. Canadian astronomer Ian Shelton was developing photographic plates and saw, to his amazement, an enormous white spot on them. He went out, raised his eyes to the sky, and noticed in the Large Magellanic Cloud a light that had not been there the night before, or earlier—the light of a supernova 170,000 light-years away. For several days it shone like several billion Suns.

The last supernova visible to the naked eye was way back in 1604. What happened that night in February? A red supergiant star, at least 10 times more massive than the Sun, was nearing its end after shining for 10 million years and then expelling its envelope to form the rings visible in the picture. Having exhausted its nuclear fuel, the star collapsed in on itself in a fraction of a second, striking its core of iron (heated to over a hundred billion degrees) and then rebounding, dispersing itself into space. The body located in the center is a neutron star about 6 miles (10 km) in diameter, so dense that a piece of its material the size of a sugar lump would weigh four hundred billion tons. Its surface is perfectly smooth—the smallest bump on its surface would be no larger than a millionth of a millimeter.

The expelled gaseous matter is still located inside the small ring, but a spherical shockwave caused by the explosion is rapidly moving toward it. It will reduce the ring to pieces before illuminating the surrounding gas like flamboyant fireworks, as can be seen in the following images of supernova remnants captured several centuries or even millennia after their explosion.

TECHNICAL DATA

The image is a composite of three colors from several exposures taken with the WFPC2 of the Hubble Space Telescope in September 1994, February 1996 and July 1997.

Bubble Nebula, NGC 7635

Cassiopeia tempts us with the striking spectacle of an open cluster containing 120 faint stars (top left of first image), a red emission nebula and a surprising bubble of ionized gaseous material. This scene extends over more than 1 degree of the sky, or more than 10 times the diameter of the full Moon. The first two pictures reproduce the natural colors of the bodies; the third, an extreme close-up of a portion of the bubble, is in coded colors. To obtain it, researchers combined a series of images taken by the Hubble Space Telescope through special filters in order to bring out the structure of this mysterious object.

The bubble is linked with the star BD+602522, which is visible at the bottom of the third, high-resolution image. This star's novelty is that it emits an extremely strong stellar wind of 4.3 million miles per hour (7 million kph). As it moves outward, this wind pushes back the ambient interstellar gases, creating a spherical shockwave—the shell of the famous bubble. It is constantly expanding, and today has reached 10 light-years. It shines by fluorescence, the effect of ultraviolet radiation emitted by the star.

Unlike a soap bubble, the surface of this cosmic sphere is far from smooth. Its wavy appearance is linked to density differences in the cold gases in the path of the wave. The knot of dense gas at the bottom right of the third image is still outside the bubble. Other, similar structures are located at the top of the image. These tortured structures are clumps of cold, dense gas, similar to the columns in the Eagle Nebula (see pages 86 and 87). The bubble is brighter on the left because it is closer to the star.

TECHNICAL DATA

Image 1: A combination of CCD images taken by the Burrell Schmidt Telescope at the Kitt Peak National Observatory, Tucson, Arizona.

Image 3: Taken by the Wide Field and Planetary Camera 2 of the Hubble Space Telescope, September 6, 1996. Color-coded in red (F658N) for doubly ionized nitrogen, green (F656N) for hydrogen alpha, and blue (F502N) for triply ionized oxygen.

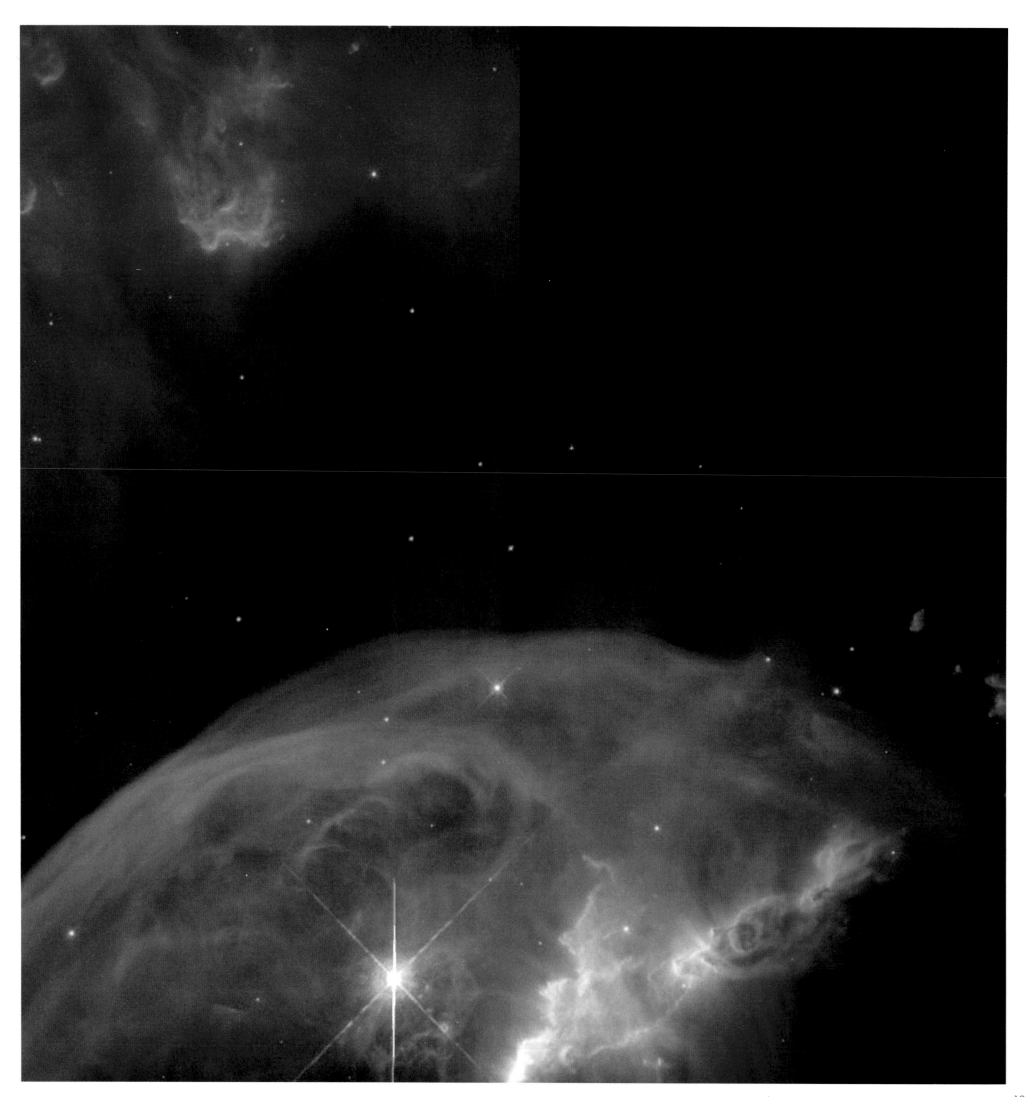

CRAB NEBULA, MESSIER 1, IN THE CONSTELLATION TAURUS

These three images reveal what happened to a star that exploded as a supernova on July 4, 1054, after shining for 50 million years. This cataclysmic explosion was doubtless observed by several civilizations of the time, including Chinese astronomers, who described a "guest star" as brilliant as the full Moon, even during the daytime, and lasting 23 consecutive days.

The first picture shows the gaseous remnants of the star that were blown into space. After being dispersed for 10 centuries, they form a turbulent and filamentary nebula colored red by ionized hydrogen. The nebula is expanding at a speed of 1,120 miles (1,800 km) per second. The center of the object is bathed in a ghostly blue light.

At the point of origin of this nebula is the corpse of a star. Though formerly as massive as 10 Suns, it crumpled into an extraordinarily dense body at the very moment of the explosion. Its mass, today equivalent to that of the Sun, is concentrated in a sphere the size of a small city. This energetic monster is a pulsar, a particular type of rotating neutron star. The Crab pulsar turns 30 times per second. As it spins, it sends into space a whirlpool of negatively charged particles (electrons) at a scarcely imaginable velocity—half the speed of light. When they are subjected to such movement, electrons emit large quantities of X rays, which appear here in blue. The neutron star is located at the bottom, to the right of the two stars (just to the left of the indigo-blue arc in the third photo). The electrons are being emitted from the smallest ring, which measures 1 light-year in diameter. They illuminate an enormous volume—a spiral 10 light-years in diameter—and as well form two immense jets perpendicular to the axis of the ring.

The Crab Nebula is located 6,500 light-years from the Sun.

TECHNICAL DATA

Image 1: Composite image (2048 x 2048 pixels) taken by the FORS2 instrument on the Very Large Telescope during the night of November 10, 1999. Three filters were used: 429 nm (blue), 657/150 nm (green) and 673 nm (red).

Image 2: Superposition of images of the entire nebula in the radio, visible (Hubble Space Telescope) and X-ray (Chandra satellite) portions of the spectrum.

Image 3: Superposition of optical and X-ray images of the internal region of the nebula, near the pulsar. Optical view from the Hubble Space Telescope; X-ray view from NASA's Chandra satellite, taken in 2000 and 2001.

LACEWORK NEBULA IN CYGNUS: DETAILS

These two images are close-up views of an enormous colored shell whose diameter extends 130 light-years (3 degrees of arc in the sky, or six times the diameter of the full Moon). It is a record imprinted in the interstellar medium of the explosion of a blue star in a supernova about 5,000 years ago. Our ancestors probably witnessed this phenomenon; it was such a major cataclysm that it must have been visible to the naked eye, even during the day. The explosion was 1,000,000,000,000,000,000 times more violent than the most powerful hydrogen bomb. It produced a spherical shockwave that continues to spread through space at a speed of 373,000 miles per hour (600,000 kph) as it pushes away the interstellar gases. The shockwave's impact against the gases produces turbulence. At the interface, the gases ripple in a complex arabesque of entangled and twisted filaments, such as the curls in the first image. We are able to admire this lacework owing to fluorescence of the various gases present.

The delicate ribbon appearing in the second image is an even more detailed view of the gas clouds as they are subjected to the impact of the shockwave. The structures visible here are the most detailed ever recorded within a supernova remnant. We see the layer of gas in profile, its thickness only one fifty-thousandth of the shockwave's total diameter. The undulations imitate perfectly what you see through a diver's mask at the interface between air and water in a pool or the ocean. The shockwave originates at the bottom of the image.

TECHNICAL DATA
Image 1: Recorded at the focus of the Canada-France-Hawaii Telescope.
Image 2: Obtained November 1997 using the Hubble Space Telescope's WFPC2 with an exposure of about two hours. The light recorded comes from hydrogen. The image measures 150 x 70 arc seconds.

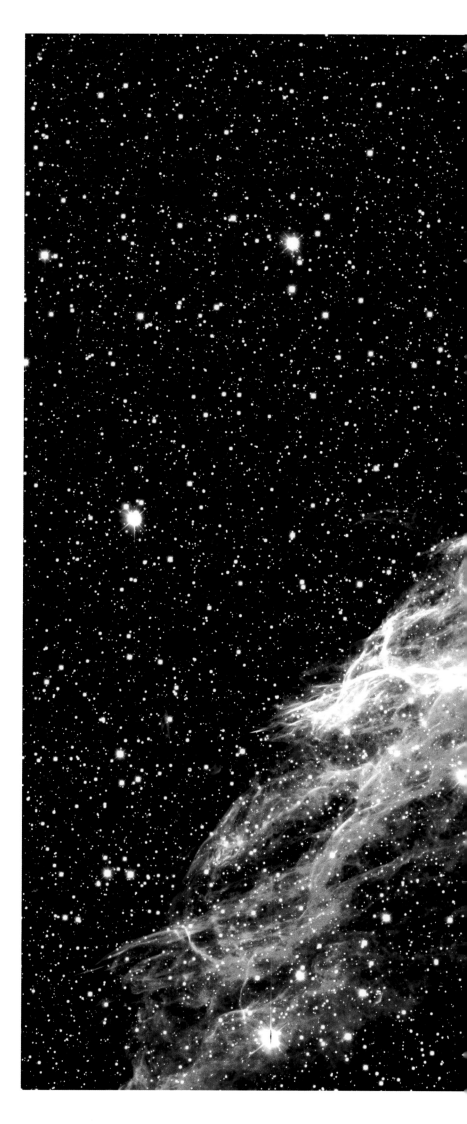

REMNANTS OF THE VEIL SUPERNOVA
(DAVID MALIN)

These shimmering curls are the last vestiges of a star in the constellation Vela (the veil) that disintegrated as a supernova 120 centuries ago. For a few days it was the brightest object in the sky, lit up like 10 billion Suns. Today only a pulsar a few miles in diameter remains of the star itself. It is located roughly in the center of the first image, but is too faint to be seen. The spherical shockwave released during that dramatic event left clearly visible evidence of its presence. While spreading through the interstellar medium at lightning speed, the shockwave pushed the star's debris ahead of it for a long distance, forming a hot, luminous wave within the sea of ambient gas. As this wave advanced, it grew larger and formed a flamboyant gaseous shell with turbulent contours, 150 light-years in diameter. Here we see two partial views. Pink filaments of hydrogen are tangled with fluorescent nitrogen shreds, whereas the colder ambient gas in front of the shock zone emits a bluish glow. Over the millennia, the gas will continue to expand, cool, fade and, finally, disperse to be reabsorbed into the ambient medium.

The matter that the deceased star expelled into the cosmos is precious. It contains heavy elements (nitrogen, carbon, oxygen, magnesium, iron, etc.) that the star forged inside itself during the main sequence. Then, during its dazzling explosion, it formed elements heavier than iron (cobalt, nickel, copper, zinc—all the way to uranium). As they were being blown away from the star, some of these atoms combined to form various molecules such as water and ammonia, as well as dust grains. Continuing their long cosmic voyage, these chemical substances will go on to enrich the interstellar clouds sown far and wide in our galaxy. With luck on their side, those clouds will condense to give birth to a new generation of stars, possibly accompanied by planets.

Remnants of the Veil Supernova

GALAXIES AS FAR AS THE EYE CAN SEE

Several billion billion billions of stars populate the observable universe. Unlike the grains of sand on a beach, stars are not uniformly distributed in the cosmos. They are grouped in gigantic structures of incomparable majesty—the galaxies. Astronomical observation has collected a veritable bestiary of galaxies, and here are the most beautiful and surprising representatives.

Galaxies gather in clusters, and clusters in turn gather to make superclusters in the form of stretched-out filaments, an immense web of three-dimensional cosmic lace. Current telescopes can capture galaxies that are extremely distant and very young—scarcely out of childhood. But their birth, shortly after the Big Bang, still remains a mystery. Did the embryonic galaxies exist as lumps within the original gaseous matter? Was the gaseous matter not homogeneous? Will we soon crack the mystery of this "age of shadows," as astrophysicists call it?

As for the real size and shape of the universe, for some researchers it may be smaller than expected. We could be living in a cosmic hall of mirrors that reflects each galaxy several times over. Other researchers support the "self-reproducing universe" theory, in which our universe may be simply a bubble created by another universe with physical properties different from ours, itself emerging from a similar entity, and so on, *ad infinitum*.

Dust filaments in galaxy NGC 891, seen edge-on

PAGE 140

Dust filaments in galaxy NGC 891, seen edge-on

This high-resolution image shows that this nearby galaxy is the site of some curious phenomena. In appearance, certainly, it resembles our Milky Way in that both of them are spiral galaxies that we see edge-on. Their spiral arms contain young blue stars and vast, dark clouds of dust—very clearly visible here—from which the stars came. These clouds seem to be sandwiched in the middle of the galactic disk as a band about 500 light-years thick. In our galaxy these clouds are easily visible to the naked eye; they darken the bright band of the Milky Way, notably in Sagittarius and Cygnus.

Recent images of NGC 891, however, reveal strange dark filaments—not present in the Milky Way—that extend perpendicular to the plane of the galaxy from the spiral arms in the direction of the halo. These extensions probably come from dark gas clouds, but what phenomenon is responsible for their formation? Have supernova explosions blasted interstellar matter into space? Or is this galaxy the site of external disturbances? The mystery remains.

TECHNICAL DATA
Image from the 11.5-foot-diameter (3.5 m) mirror of the WIYN Telescope, obtained by Blair Savage and Chris Howk, University of Wisconsin, National Optical Astronomy Observatories (NOAO).

PAGE 143

Spiral galaxy NGC 4414
(HUBBLE SPACE TELESCOPE)

This spiral galaxy, located 60 million light-years away, resembles our own galaxy, the Milky Way. Both contain about a hundred billion stars that seem to move in a vast whirlpool. This nearly face-on view clearly shows the presence of billions of mature yellow and red stars in the very dense central region of this maelstrom—the galactic bulge. Their age is estimated at over 10 billion years. This region could conceal in its core a black hole, with a mass equivalent to a million Suns.

The outer arms show two indications of the abundance of their star-formation regions. The dominant blue reveals the presence of stars formed recently—within the last hundred million years. The most brilliant can be distinguished individually in this very high-resolution image supplied by the wide-field camera of the Hubble Space Telescope. The spiral arms are also extremely rich in dark, dusty interstellar clouds, their irregular outlines standing out from the star field. These clouds constitute the primary material from which stars are formed.

TECHNICAL DATA
Image obtained by the Hubble Space Telescope during the Extragalactic Distance Scale Project. The purpose of this program is to find and study variable stars in order to precisely determine galactic distances and thus measure the rate of expansion of the universe. The observation was led by an international team directed by Wendy Freedman of the Carnegie Institute of Washington Observatory. The galaxy was observed 13 times in two months during 1999 using the WFPC2.

Tadpole Galaxy, UGC 10214
(HUBBLE SPACE TELESCOPE)

This unusual galaxy, located 420 million light-years away, has been the scene of an event of great violence. It was distorted by a smaller and very compact galaxy that crossed its disk from left to right. This intruder appears in blue, through the spiral arms of the host galaxy, at top left of this image; it is, however, now 300,000 light-years behind it. Under the impact of the blow, UGC 10214 lost some of its matter, which was cast into space as a streak of stars and gas nearly three times longer than the Milky Way.

But this collision did not have only devastating effects—the prodigious energy liberated during the galactic encounter is the origin of a bloom of stars. The blue spots visible in the streak are actually huge clusters, each containing millions of massive young stars, 10 times hotter and a million times more brilliant than our Sun. In the distant future, the streak will probably separate from its source galaxy and break up, tearing apart in the areas around the darker zones and separating the blue clusters to form dwarf galaxies orbiting their parent. Some of the star clusters will wander freely in the respective halos of these galaxies.

The background of this image—6,000 galaxies of many varied, sometimes colliding forms—is also of great interest. In some respects it is a fossil deposit from the Big Bang. Some of these galaxies are extremely distant, and they look as they did only a billion years after the beginning of cosmic expansion in the dark age of the universe's infancy, about which we know little or nothing.

TECHNICAL DATA

Image obtained using the ACS (Advanced Camera for Surveys) installed on the Hubble Space Telescope. It was taken in two exposures on April 1 and 9, 2002. The colors result from coding; they are reconstituted from three separate images taken through filters isolating near-infrared, orange and blue radiation.

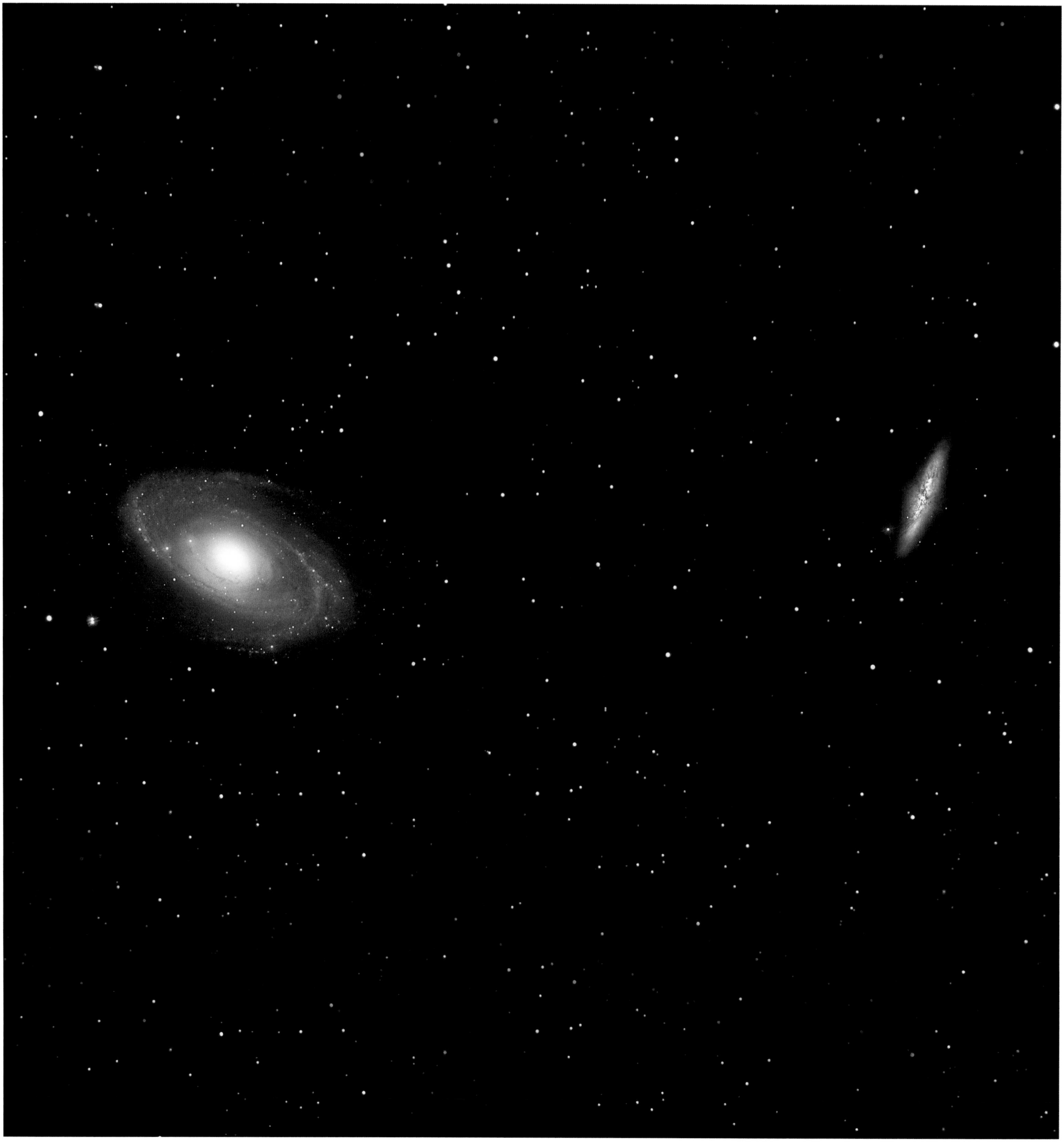

PAGE 146

Spiral galaxy NGC 300

A vast eddy of stars held together by the force of gravity, this typical spiral galaxy shimmers seven million light-years from us in the southern constellation Sculptor. Because of its short distance away (on a cosmic scale), this object seems fairly spread out at 25 arc minutes, just a little less than the full Moon. It is easily visible through binoculars.

For its observers, NGC 300 has the advantage of appearing face on, unlike our galaxy, the Milky Way, which is seen edge-on (the Solar System is actually located in the galactic plane). If we could gaze upon the Milky Way from outside, it would certainly provide us the same spectacle. The central bulge is rich in very long-lived yellow stars. The blue giant stars, born at the same time as the yellow ones, at the beginning of the universe, are already dead and have not been replaced. In the spiral arms, on the other hand, generations of all types of stars (red, yellow, white and blue) have followed one after another since the beginning. The yellow and red stars, the most abundant, are also the smallest, and their light is masked by the blinding glare of blue giants.

The red glow scattered throughout the spiral arms is from emission nebulas, vast interstellar clouds of gas and dust excited by the stars being born in their interior or wandering nearby.

TECHNICAL DATA
Data collected by the 7-foot-diameter (2.2 m) MPG/ESO telescope of the European observatory at La Silla, Chile, using the Wide Field Imager (WFI) CCD camera (67 million pixels).

PAGE 147

Galaxies Messier 81 and Messier 82

Twelve million light-years away, in the constellation Ursa Major, wanders a family of 13 interacting galaxies, including the giant spiral galaxy M81 in the picture's foreground. It is a favorite of amateur astronomers because its great brightness makes it easy to observe, even with a small instrument. Some people have even seen it with the naked eye during exceptionally dark nights.

For very different reasons, professional astronomers are interested in the pair that M81 forms with its neighbor, the irregular galaxy M82. This research involves reconstructing a truly cosmic drama—the two galaxies collided and then moved apart, reaching the distance of 150,000 light-years that currently separates them. The two galaxies bear the marks of this encounter. They shine even brighter because of the surge in stellar births caused by their tidal effects on each other. As well, linear filaments are being drawn out in the central region of M82, which is subject to more significant disruptions because of its lower mass.

The Hubble Space Telescope detected within M82 at least a hundred very brilliant, newly formed globular clusters, each containing 100,000 stars. By studying their stellar composition, astronomers have deduced their age. From that they have estimated that the two galaxies interacted 600 million years ago, and that the interaction itself would have lasted a few hundred million years.

PAGE 149

Jet from elliptical galaxy Messier 87 / NGC 4486

This immense jet of exotic matter, 5,000 light-years long, is composed of electrons and other subatomic particles shooting into the cosmic vacuum at close to the speed of light (186,000 miles / 300,000 km per second). It is accompanied by enormous bursts of X-rays, ultraviolet radiation, gamma rays and other forms of radiation.

Astronomers had already noticed in 1918 a straight "ray" coming from a region in Virgo. In the 1950s it was also noticed that the area was one of the most powerful radio sources in the sky. Several decades were needed to prove that this enormous energy was coming from the center of one of the most massive elliptical galaxies known. That galaxy, located 50 million light-years away, contains more than 10 trillion stars. This image does not make it possible to tell them apart, but globular clusters, each containing several hundreds of thousands of stars, can be distinguished as bright points. It is estimated that the galaxy contains 15,000 clusters in total.

The source of this terrifying power is probably a super-massive black hole hidden in the center of the galaxy. This black hole would have already swallowed up a mass equivalent to two or three billion solar masses. Irresistibly it pulls in stars, gases and dust. During its descent toward that voracious mouth, matter begins to rotate around it at a speed of 1.2 billion miles per hour (2 billion kph), ionizing and heating up to several million degrees. The electrons freed by the ionization follow the black hole's intense and entangled magnetic field lines. Spiraling along the lines, they emit synchrotron radiation tangentially to their path—the famous jet.

M87 is far from being the only galaxy whose source of energy is most likely a black hole. Thousands of other so-called active galaxies have already been counted. But the Milky Way and other ordinary spiral galaxies may also conceal a black hole at their center. These, although monstrous on the scale of the Solar System, nonetheless pale in comparison with M87's, which may be a thousand times more massive.

TECHNICAL DATA
Image obtained from data collected in 1998 by the Hubble Space Telescope's WFPC2 by combining four exposures in ultraviolet, blue, green and infrared light.

Sombrero Galaxy, Messier 104
(VERY LARGE TELESCOPE)

In the constellation Virgo, about 50 million light-years from Earth, wanders one of the most beautiful galaxies known. Numerous, more distant galaxies shine timidly through the thin texture of its halo.

This gigantic spiral galaxy contains several trillion stars. Although photographed repeatedly, it had never revealed such fine detail as in this view taken by the European Very Large Telescope, located in Chile. The resolution of the image is 170 light-years. This picture reveals for the first time the complex structures of the dark dust clouds laid out in equatorial bands, the very prominent bulge, and the diffuse halo sprinkled with many small luminous sources—globular clusters. But there is no sign of blue giant stars or the nebulas from which they normally originate; only low-mass, very old yellow stars populate this galaxy. Has the formation of new stars stopped? Did this galaxy transform all of its interstellar gas into stars all at once, right at the beginning of its life or during a collision with another galaxy? Or perhaps the dark bands conceal stellar nurseries? Further, why is the core of this galaxy a source of intense radiation in all wavelengths? Does it conceal a black hole weighing about a hundred solar masses? The mystery remains unsolved.

TECHNICAL DATA

Image obtained using the FORS1 instrument of the Very Large Telescope's third dome, which was still incomplete, on January 30, 2000. It is a composite of three CCD images taken at different wavelengths: 554 nm (exposure of 120 seconds) in blue; 657 nm (exposure of 120 seconds) in green; 768 nm (exposure of 240 seconds) in red.

NEEDLE GALAXY, NGC 4565

The constellation Coma Berenices hosts this very beautiful spiral galaxy, discovered in 1785 by William Herschel and visible edge-on, like the Milky Way. It shines 31 million light-years from us. It is often photographed by amateur astrophotographers because of its great brightness. This high-resolution view was taken by one of the most powerful instruments in the world, the Canada-France-Hawaii Telescope, which is located at an altitude of 13,800 feet (4,200 m) on the summit of Mauna Kea, an extinct volcano in the Hawaiian Islands. This 12-foot-diameter (3.6 m) telescope is a cooperative project of the University of Hawaii and various Canadian and French astronomical institutions. The instrument is located a few hundred yards from NASA's 10-foot (3 m) telescope, the twin domes of the Keck Telescope, which has the largest mirrors in the world (33 feet / 10 m in diameter), and finally the 27-foot (8.3 m) Japanese telescope, Subaru. The site is one of the highest used for astronomy, and the nights are exceptionally clear. The other high place for astronomy is in the Cordilleras of the Chilean Andes, where several telescopes are located, both American (Las Campanas) and European (La Silla and Paranal, site of the four domes of the Very Large Telescope).

COLLIDING GALAXIES NGC 4676

In April 2001 the Hubble Space Telescope's ACS camera captured this strange galactic object 300 million light-years away, in the constellation Coma Berenices. It involves a couple of spiral galaxies named "The Mice," because of the two long tails of stars and gas emanating from them.

The blue areas visible in the galaxy at the left and in the long streak at the right are particularly massive stellar clusters. Their mass is comparable to that of dwarf galaxies such as the Magellanic Clouds, which orbit our own galaxy. They sprang from clouds of gas disturbed by the titanic forces of gravitational attraction that each galaxy exerts on its neighbor.

Computer simulations were developed to test various scenarios for the formation of this two-headed object. They suggest that the two galaxies collided 160 million years ago and will eventually fuse to form a single elliptical galaxy. It is not known what will happen to the gas and stars forming the tails. Will they be swallowed by the galaxy or will they go into orbit around its halo?

The spectacle of this cosmic collision previews what might happen if our galaxy gets too close to one of its large neighbors, the Andromeda galaxy (M31), which is rushing toward it. They could collide in the very distant future—that is, in several billion years. The Sun will certainly be dark by then.

TECHNICAL DATA

Image obtained by combining three separate images taken April 7, 2001, through blue, orange and near-infrared filters on the Hubble Space Telescope's ACS camera.

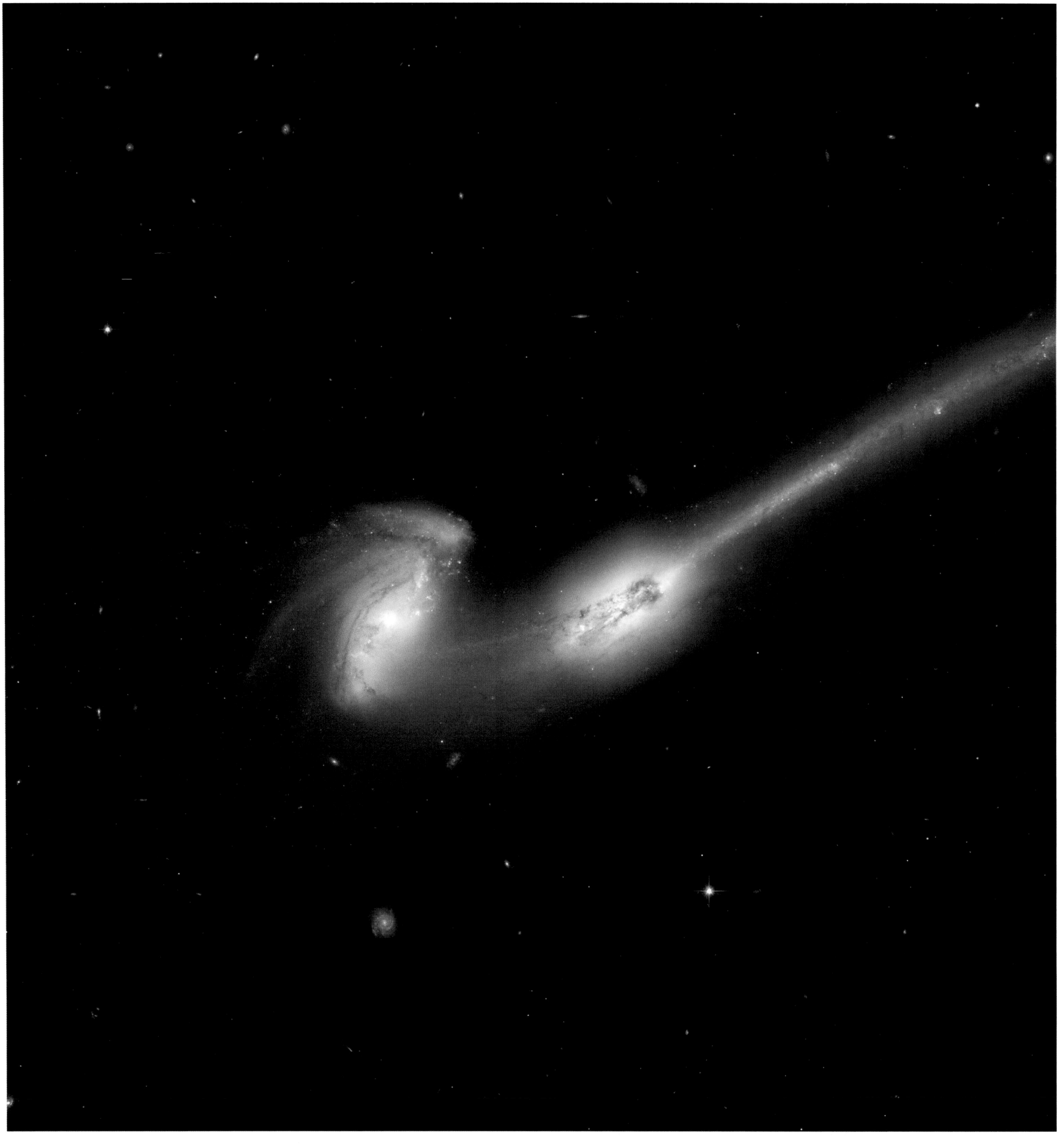

Galaxy NGC 4319 and quasar Markarian 205

(HUBBLE SPACE TELESCOPE)

This image shows an unusual pair of bodies. The majestic whirlpool of the spiral galaxy NGC 4319 competes with the quasar cataloged as Markarian 205 (top right of the picture). The two objects are extremely far away from each other. NGC 4319 is 80 million light-years from Earth, whereas Mrk 205 is a billion light-years away. The light that we receive from the two objects has traveled 80 billion and one billion years respectively before reaching us, so the apparent alignment of the two objects is only the result of chance.

They each display unusual properties. The bands of dust in the spiral arms of NGC 4319 are abnormally dark. As well, in the picture's bottom left an arm very discreetly reveals its unexpected presence in the area. These two properties suggest that this galaxy has in the past undergone significant gravitational disruption. Probably it interacted with one of its nearby neighbors, the galaxy NGC 4291, which is not visible in this image.

Mrk 205 is a relatively close quasar (from "quasi-stellar objects"); most of these bodies are much farther away. When the first quasars were discovered, they appeared to be point sources like stars, but their extraordinarily high luminosity—like several billion stars—left observers perplexed. When powerful telescopes became available, astrophysicists were able to show that a quasar was not a star but the central region of a galaxy with a particularly brilliant bulge, within which was probably concealed a black hole the size of our Solar System. As a black hole swallows celestial matter, the material falling into it releases a considerable amount of light, radio, UV and X-ray energy. The center of the galaxy alone shines a hundred times brighter than our entire Milky Way. In this picture, despite the blinding brightness of the quasar, a weak galactic halo can be seen. It is deformed by the presence of a compact galaxy located just below it.

TECHNICAL DATA

Composite image created from archived data for NGC 4319 (March 1997) and exposures taken with the Hubble Space Telescope's WFPC2 in February 2002, with three filters.

HOAG'S OBJECT: A RING GALAXY
(HUBBLE SPACE TELESCOPE)

This image is the most accurate ever obtained of this exotic and rare celestial object located in the constellation Serpens. It was discovered in 1950 by the astronomer Art Hoag, who at the time had only a very fuzzy image, not at all comparable to this one. He mistook it for a planetary nebula before realizing that it could be a galaxy. But it took two decades to confirm that Hoag's Object really was a galaxy. The object, which is 600 million light-years away, has a diameter of 120,000 light-years. The bluish ring gets its color from a rich collection of massive, brilliant young stars—blue supergiants. Their abundance signifies a veritable outburst of simultaneous stellar births caused by contraction of interstellar gases. The yellow tint of the central nucleus reveals the presence of much older stars. The dark part separating the two structures likely contains stellar clusters that are too faint to be visible. Surprisingly, a body resembling a miniature version of Hoag's Object can be found in this dark zone, at the top and a little to the right. It is probably another ring galaxy farther away.

In total, only a few dozen ring galaxies are known. One scenario to explain their formation involves a collision between a large disk-shaped galaxy and a small elliptical galaxy rushing perpendicularly toward it. Just the same, no intruding galaxy can be detected in either Hoag's Object or its vicinity. Another hypothesis under consideration is that the yellow central galaxy could have captured, and pulled into orbit around it, shreds of stellar material from a second galaxy that came too close to it. This encounter would have happened two or three billion years ago.

TECHNICAL DATA

Image obtained with the Hubble Space Telescope's WFPC2 on July 9, 2001, with an exposure of 3.6 hours, through three filters.

SEYFERT'S SEXTET GALAXY CLUSTER
(HUBBLE SPACE TELESCOPE)

Astronomers continue to build up evidence that galaxies interact through turbulent, cataclysmic collisions that radically modify their structures. This picture of a region in the constellation Serpens is one of the most remarkable examples. To understand the scene shown here, it helps to first mentally remove the small spiral galaxy in the center of the image, as it is located in the background and does not interact with the others. The object at bottom right is not a galaxy. Carl Seyfert, the astronomer who discovered this cluster at the end of the 1940s, already suspected, despite the very low resolution of images at the time, that this glow was an immense trail of stars.

The four other bodies, all located 150 million light-years from Earth, span distances of scarcely 100,000 light-years in a volume less than that occupied by the Milky Way. They are so close together that gravitational forces are deforming them, ripping certain areas apart. They are struggling against death and mutual annihilation. Three galaxies, the elliptical one second from the top and the two spiral galaxies at the bottom, are being distorted by the effect of tidal forces. Stars have been torn away from them, forming the halos that surround them. The galaxy at bottom center has lost an entire stellar layer, which is escaping in the form of an immense trail of stars 35,000 light-years long. In comparison, the small spiral galaxy seen edge-on at top center seems to have suffered little damage; its stars remain within the limits of the galaxy. All the same, its disk is slightly curved.

In contrast with most galactic interactions that have been observed at high resolution, this one does not seem to have caused the formation of clusters of blue stars—perhaps the sextet is only beginning its interaction. Over the coming hundreds of millions of years the four galaxies may, after much tumult, coalesce to form a single elliptical or irregular galaxy. Today astronomers think that many elliptical galaxies, maybe even all of them, resulted from coalescence. This must have been especially frequent at the beginning of the universe, when it was smaller and denser than it is now, increasing the odds of collisions.

TECHNICAL DATA

Image obtained on June 26, 2000, by the Hubble Space Telescope's WFPC2. Length of exposure: 4.1 hours. Filters used: ultraviolet, blue, V 555 and infrared.

PAGE 163

GRAVITATION LENSES IN ABELL 1689 GALACTIC CLUSTER
(HUBBLE SPACE TELESCOPE)

In this image, a group of yellow galaxies accompanies red and blue circular arcs. Such a spectacle has been observable only since the invention of powerful astronomical instruments in the 1990s. The presence of the arcs confirms Albert Einstein's theory of general relativity. The scientist had predicted that light rays are bent during their passage through areas where space-time is deformed by enormous masses. These act as virtual lenses, modifying the appearance of objects behind them.

Appropriately, the yellow galaxies in the foreground are part of one of the most massive galactic clusters known. It encompasses trillions of stars, generating an enormous gravitational field. This increases the bending of light coming from galaxies located behind it, resulting in their arc-shaped appearance.

This exceptional image reveals 10 times more arcs than would a terrestrial telescope pointed in the same direction. The faintest galaxies are probably more than 13 billion light-years away from us. They capture the interest of researchers because of their very young age.

The study of gravitational lenses is also crucial for trying to understand how unseen dark matter, which seems to make up over 90 percent of the mass in the universe, is distributed. The quantity of matter, both luminous and dark, contained in this cluster can actually be determined by studying the shapes of the arcs.

TECHNICAL DATA

Composite image acquired in visible light and infrared (representative colors) obtained with the Hubble Space Telescope's ACS camera in June 2002. It required a 13-hour exposure.

GLOSSARY

A

ASTEROID: A rocky or metallic chunk orbiting the Sun. If disrupted by a chaotic movement, it could collide with a planet or satellite.

ASTEROID BELT: A collection of millions of asteroids orbiting between Mars and Jupiter.

ASTRONOMICAL UNIT (AU): The unit of measure-ment for short astronomical distances. One AU is equivalent to the average distance between the Sun and Earth, about 93 million miles (150 million km).

ATMOSPHERE: The layers of gas surrounding a planet or moon.

B

BIG BANG: A mocking term coined in the 1950s by the British astronomer Fred Hoyle, who was opposed to the idea of an explosion as the origin of the universe. Today it designates the accepted theory, according to which the cosmos was extremely dense at its beginning and has been continuing to expand and cool for 15 billion years.

BINARY SYSTEM (DOUBLE STAR): Two stars orbiting a common center.

BLACK DWARF: The last stage in the evolution of a star, when its mass is less than 1.5 solar masses and it no longer radiates light.

BLACK HOLE: A superdense object (the corpse of a very large star or core of a galaxy) corresponding to a region of space-time deformed to such an extreme extent that it swallows up everything that enters it, without possibility of return—even light.

BLUE GIANT STAR: A young star more massive, hotter and brighter than the Sun, but with a short life expectancy (about 30 million years) and destined to end its life as a supernova.

BLUE SUPERGIANT STAR: A monstrous star as brilliant as 10,000 or more Suns.

BODY: Any celestial object, whatever its nature—planet, star, comet, nebula or galaxy.

BROWN DWARF: A body not quite large enough to start nuclear fusion like a star, with a diameter scarcely larger than one one-hundredth of the Sun's.

BULGE: The central bright, spherical region of a spiral galaxy, which is populated with old stars.

C

CARBON DIOXIDE (CO_2): A gas composed of two atoms of oxygen and one of carbon, found in the primitive atmosphere of some planets. It helps in determining the atmosphere's temperature (see also Greenhouse Effect).

COMET: A block of ice and dust orbiting the Sun.

CONSTELLATIONS: Groups of apparently close stars, symbolically linked together since antiquity as depicting mythological heroes or animals in the sky. Also, the 88 regions of the sky established by the International Astronomical Union in 1922, using the ancient constellations as reference points.

COSMIC EXPANSION: The expansion of space in space-time, carrying along the galaxies, which are separating from each other as the Big Bang progresses. This separation shows up as a shift toward red in the light that they transmit to us. The most distant are moving away the fastest, and measurement of their redshift helps in estimating their distance.

COSMOLOGY: The study of the evolution and structure of the universe on a very large scale.

D

DARK MATTER: See Missing Mass

DARK NEBULA: An interstellar cloud of gas and dust that is opaque to visible light, concealing the celestial objects behind it.

DIFFUSE NEBULA: A colored interstellar cloud.

DOUBLE STAR: See Binary System.

E

ELECTROMAGNETIC RADIATION: Visible light, infrared waves, microwaves and radio waves, which are all less energetic, and ultraviolet rays, X rays and gamma rays, which are more energetic.

ELLIPTICAL GALAXY: A galaxy whose form may range from spherical to cigar-shaped.

EMISSION NEBULA: An interstellar cloud whose gases are excited by ultraviolet radiation from nearby stars, leading to emission of radiation through fluorescence, as in neon lights.

EUROPEAN SOUTHERN OBSERVATORY (ESO): A European consortium that operates several telescopes in Chile, including the four domes of the VLT (Very Large Telescope), located on Mount Paranal—currently the largest instrument of its type.

EXOPLANETS (EXTRASOLAR PLANETS): Planets that orbit stars other than the Sun, discovered using Earth-based telescopes starting in 1995.

F

FLUORESCENCE: The emission of colored radiation by atoms when, after being excited (for example, by ultraviolet radiation), they return to a lower energy state, as in neon lights. Each atomic element radiates at one or more characteristic wavelengths. Spectroscopic measurement of these wavelengths reveals the elements present in the radiating objects.

FUSION (THERMONUCLEAR): The union of two atomic nuclei to form the nucleus of a different element, for example, helium from hydrogen or carbon from helium. It happens at very high temperatures and pressures, such as in the core of a star, and is accompanied by a significant release of energy.

G

GALACTIC HALO: A vast spherical region surrounding a spiral galaxy. It contains very few stars except for globular clusters as old as the galaxy itself.

GALAXY: A vast system containing between 10 million and 100 trillion stars held together by their mutual gravitational attraction.

GALAXY CLUSTER: A group of galaxies that can contain between 10 and several thousand galaxies, held together by their mutual gravitational attraction.

GLOBULAR CLUSTER: A dense spherical group of several thousands to millions of stars bound by their gravitational attraction. Very old clusters—12 billion years old—are found in the spherical halo of spiral galaxies (see also Galactic Halo).

GRAVITATIONAL FIELD: A field that exerts an attractive force on bodies. The closer or more massive the bodies, the stronger the force. Also, deformation of space-time.

GREENHOUSE EFFECT: The trapping within a planetary atmosphere of infrared rays, which transport heat, by certain gases, including water vapor, carbon dioxide and methane.

H

HEAVY ELEMENTS: Atoms heavier than hydrogen and helium, up to uranium. They are formed from light elements by nuclear reactions in stars.

HELIUM: The second simplest and most abundant element in the universe, after hydrogen (see also Hydrogen).

HUBBLE SPACE TELESCOPE: A telescope with an 8-foot-diameter (2.4 m) mirror that was placed in Earth orbit by the American Space Shuttle *Discovery* in April 1990. Its dimensions: 42.6 feet (13 m) long and 16 feet (5 m) wide, with a 39-foot (12 m) wingspan and a mass of 12 tons. It is named after one of the discoverers of the expansion of the universe, Edwin Hubble. It can resolve details 10 times finer than the best Earth-based telescopes and can reveal five billion bodies too faint to be seen with the naked eye.

HYDROGEN: The simplest, lightest, most stable and most abundant element in the cosmos, formed at the beginning of the universe.

I

INTERSTELLAR MEDIUM: The space between the stars and the material in it. Its average temperature is about –450°F (–270°C). It is not empty, but contains atomic and molecular gases and dust.

IONIZED MATTER: Atoms or molecules that have been electrically charged through the loss or gain of one or more electrons. Ionized matter is abundant in nebulas (see also Plasma).

IRREGULAR GALAXY: A galaxy with no particular shape that is a site of many star births.

K

KUIPER BELT: A belt of millions of comets orbiting the Sun at a distance between 500 and 1,000 astronomical units.

L

LIGHT ELEMENTS: The simplest atomic elements, hydrogen and helium, which were formed at the beginning of the universe.

LIGHT-YEAR: The principal unit of astronomical measurement, the distance traveled by light in one year. Light spreads at about 186,000 miles (300,000 km) per second, so one light-year is 6,000,000,000,000 miles (9,460,000,000,000 km). The distance to the Sun is 8 light-minutes, the closest star to it is 4 light-years away, and the closest galaxy is 170,000 light-years distant. If we look at a galaxy 60 million light-years from Earth, we see it as it was 60 million years ago—at the time of the extinction of the dinosaurs.

LOCAL CLUSTER: The cluster of galaxies containing the Milky Way and its nearest neighbors, including the Large and Small Magellanic Clouds and the Andromeda galaxy, which is 2.3 million light-years from Earth.

LOCAL GROUP: The group of some 30 galaxies to which the Milky Way belongs.

M

MAGELLANIC CLOUDS: A pair of dwarf galaxies orbiting the Milky Way 170,000 light-years away. Visible only from the Southern Hemisphere, they are named after Ferdinand Magellan, the navigator who discovered them.

MAGNETIC FIELD: A field that exerts force on electrically charged particles or magnetic materials.

MAIN SEQUENCE: The stage of development of 90 percent of the stars—those which, like the Sun, shine without difficulty, drawing their energy from the nuclear fusion of hydrogen.

MESSIER CATALOG: A list of celestial objects (galaxies, nebulas, etc.) observed by the French astronomer Charles Messier during the 18th century.

METEORITE: A metallic, rocky or mixed fragment of an asteroid or, less often, of a planet.

MILKY WAY: The spiral galaxy that contains the Sun. It got its name because of its milky appearance when seen across the dome of the sky.

MISSING MASS (DARK MATTER): An invisible substance of unknown nature whose existence is deduced from its gravitational effects, especially on galaxies and galaxy clusters.

MOLECULAR CLOUD: A vast collection of interstellar gas and dust containing various organic and inorganic molecules, some of them complex.

MOON: A natural satellite; a rocky body orbiting a planet.

N

NEBULA: Historically, any celestial object with a diffuse appearance, such as a galaxy, interstellar cloud, etc.

NEUTRON STAR: The remnants of a super-massive star that has exploded in a supernova. These bodies are hyperdense and compact— with a diameter of only 6 miles (10 km), their mass is greater than that of the Sun. They are composed only of neutrons, which are neutral particles from atomic nuclei.

NGC (NEW GENERAL CATALOG): A catalog of celestial bodies compiled in the 19th century.

O

OBSERVABLE UNIVERSE: A celestial sphere about 15 billion light-years in diameter that contains the visible celestial objects. When we see them, they are not all the same age. The more distant are older because of the finite speed of light.

OORT CLOUD: A collection of cometary nuclei orbiting at the edge of the Solar System, named after 20th-century Dutch astronomer Jan Hendrik Oort.

OPEN CLUSTER: A group of several tens to many thousands of relatively young stars born from the collapse of a single cloud. They tend to escape to spread out to the four corners of their galaxy.

ORBIT: The circular or elliptical trajectory of one body around another, generally more massive, body.

P

PLANET: A large sphere orbiting a star. Incapable of existing in a free state and not producing any light itself, a planet shines by reflecting light from its star. Some planets have a rocky surface like Earth's, while other planets are gas giants.

PLANETARY NEBULA: Gaseous debris expanding from a star of less than 1.5 solar masses.

PLASMA: The fourth state of matter, the most abundant in the universe. It is gas of variable density in which all the atoms have been ionized, and is the matter that makes up stars, stellar winds and colored nebulas.

PLATE TECTONICS: Movements of Earth's crust, giving rise to oceans and mountain chains such as the Himalayas and the Alps.

PRIMORDIAL GALAXY: An extremely distant galaxy (12 or 13 billion light-years), observed while it is still forming.

PULSAR: A neutron star that rotates extremely rapidly and emits bursts of electromagnetic radiation (visible light, radio waves, X rays, etc.) at regular intervals, like a lighthouse. Its density can reach a hundred billion kilograms per cubic centimeter.

Q

QUASAR: The extremely brilliant center of an active galaxy that probably conceals a giant black hole.

R

RADIO ASTRONOMY: The study of radio waves emitted by celestial bodies such as pulsars, galaxies or even matter that appeared shortly after the Big Bang.

RED DWARF: A star with a relatively cold surface, about 10 times less massive than the Sun and a thousand times less brilliant, but with a very significant life expectancy—at least 200 billion years, and up to a trillion years. Such a star's life will end quietly as it expires in the form of a black dwarf.

RED GIANT STAR: A star that has expanded and cooled, the source of a planetary nebula, and destined to collapse into a white dwarf. A red giant represents an advanced stage in the development of stars like the Sun.

RED SUPERGIANT STAR: A massive star at the end of its life that could explode into a supernova at any moment.

REFLECTION NEBULA: An interstellar cloud of gas and dust reflecting the light emitted by nearby stars.

REVOLUTION: The movement of one celestial object in a circular—or, more often, elliptical—path around another.

RING (PLANETARY): A belt made up of rock or ice chunks and dust around a planet.

S

Satellite: See Moon.

SHOCKWAVE: A region where interstellar gas is compressed, heated or excited under pressure from a flow of particles. It can be compared to the pressure wave from an explosion or earthquake.

SOLAR SYSTEM: The Sun and all the bodies that orbit it (planets, moons, comets, asteroids, dust, etc.).

SPACE-TIME: Four-dimensional geometric coordinates postulated by Jules Poincaré and Hermann Minkowski and then taken up by Albert Einstein.

SPECTRUM: All the radiation emitted or absorbed by an object, obtained by breaking up its light with; for example, a prism.

SPIRAL ARM: A dense region of a spiral galaxy, the site of star births.

SPIRAL GALAXY: A galaxy whose stars and interstellar gases are essentially distributed along spiral arms winding around a central spherical bulge.

STARS: VERY MASSIVE, incandescent gaseous spheres composed mostly of hydrogen and helium. The gas is maintained in a plasma state by nuclear fusion reactions taking place in the star's core. Their dimensions range from one one-hundredth to a hundred times the size of the Sun.

STELLAR WIND (SOLAR WIND): The flow of energetic radiation (UV and gamma) and charged particles continuously emitted by a star.

SUPERCLUSTER: A group of tens of thousands of galaxies brought together by gravitational attraction in the form of a pancake 90 million light-years in diameter.

SUPERNOVA: The titanic explosion of a large star that has run out of nuclear fuel. Most of the stellar matter is expelled into the interstellar medium.

U

ULTRAVIOLET RADIATION (UV): Electromagnetic radiation with shorter wavelengths and higher energy than violet light. The most energetic ultraviolet can ionize matter.

V

VERY LARGE TELESCOPE (VLT): The most powerful group of telescopes in the world, installed at the summit of Mount Paranal (altitude 8,645 feet / 2,635 m) in Chile's Atacama Desert. Its four 27-foot-diameter (8.2 m) telescopes are so powerful that they can photograph an astronaut on the surface of the Moon. The VLT is managed by the European Southern Observatory (ESO), a consortium of nine European countries, including France.

W

WHITE DWARF: The residue from the death of the star of less than 1.5 solar masses, with progressively diminishing luminosity. This body is so dense that a thimbleful of its matter would have a mass of about 2 tons.

SUGGESTED READING

Audouze, Jean, and Guy Israel, eds. *Cambridge Atlas of Astronomy*, 3rd edition. Cambridge University Press, 1994.

Bone, Neil. *The Aurora*, 2nd edition. Chichester, UK: Wiley-Praxis, 1996.

———. *Observer's Handbook: Meteors.* London: George Philip; Cambridge, MA: Sky Publishing, 1993.

Brunier, Serge. *The Great Atlas of the Stars.* Toronto & Buffalo, NY: Firefly Books, 2001.

———. *Majestic Universe: Views from Here to Infinity.* Cambridge University Press, 1999.

———. *Solar System Voyage.* Translated by Storm Dunlop. Cambridge University Press, 2002.

Burnham, Robert. *Burnham's Celestial Handbook.* 3 vols. New York: Dover Publications, 1978.

Chartrand, Mark R. *Collins Guide to the Night Sky.* London: HarperCollins, 1999.

———. *Field Guide to the Night Sky.* New York: Knopf, 1991.

Cook, J., ed. *The Hatfield Photographic Lunar Atlas.* Heidelberg: Springer-Verlag, 1999.

Dickinson, Terence, and Alan Dyer. *The Backyard Astronomer's Guide.* Toronto: Firefly Books, 2002.

Dickinson, Terence. *NightWatch.* Toronto: Firefly Books, 1998.

Dunlop, Storm, and Wil Tirion. *How to Identify: Night Sky.* London: HarperCollins, 2002.

Fischer, Daniel, and Hilmar Duerbeck. *Hubble: A New Window to the Universe.* New York: Copernicus Books, 1996.

Freedman, Roger A., and William J. Kaufmann. *Universe.* New York: W.H. Freeman, 2001.

Harrington, Philip S. *Touring the Universe through Binoculars.* New York: Wiley, 1990.

Harrison, Edward. *Darkness at Night: A Riddle of the Universe.* Cambridge, MA: Harvard University Press, 1987.

———. *The Universe.* 2nd edition. Cambridge University Press, 2000.

Hawking, Stephen. *The Illustrated A Brief History of Time.* New York: Bantam Books, 1989.

Hoskin, M., ed. *Cambridge Illustrated History: Astronomy.* Cambridge University Press, 1997.

Illingworth, Valerie, and John O.E. Clark, eds. *Collins Dictionary of Astronomy,* 2nd edition. London: HarperCollins, 2000.

———. *Facts on File Dictionary of Astronomy*, 4th edition. New York: Facts on File, 2000.

Karkoschka, E. *The Observer's Sky Atlas*, 2nd edition. New York: Springer-Verlag, 1999.

Koestler, Arthur. *The Sleepwalkers: A History of Man's Changing Vision of the Universe.* Pelican Books: 1968.

Lachièze-Rey, Marc. *Introduction to Cosmology: A First Course.* Cambridge University Press, 1999.

Luminet, Jean-Pierre. *Black Holes.* Cambridge University Press, 1992.

Maran, Stephen P. *Astronomy for Dummies.* New York: IDG, 1999.

Mobberley, M. *Astronomical Equipment for Amateurs.* Heidelberg: Springer-Verlag, 1995.

Moore, Patrick, ed. *Astronomy Encyclopedia.* London: George Philip, 2002.

———. *Exploring the Night Sky with Binoculars*, 4th edition. Cambridge University Press, 2000.

———. *The Observational Amateur Astronomer.* Heidelberg: Springer-Verlag, 1995.

Pirani, Felix, and Christine Roche. *The Universe for Beginners.* London: Icon Books, 1997.

Reeves, Hubert, and Donald Winkler. *Latest News from the Cosmos: Towards the First Second.* Toronto: Stoddart Publishing, 1997.

Ridpath, Ian, ed. *Norton's Star Atlas*, 19th edition. London: Longman, 1998.

———. *Oxford Dictionary of Astronomy*, 2nd edition. Oxford University Press, 2003.

Ridpath, Ian, and Wil Tirion. *Collins Pocket Guide: Stars and Planets.* London: HarperCollins, 2000.

Rükl, Antonín. *Hamlyn Atlas of the Moon*, London: Hamlyn; Milwaukee: Astro Media, 1990.

Scagell, Robin. *Philip's Stargazing with a Telescope.* London: George Philip, 2000.

Spence, Pam. *The Universe Revealed.* Cambridge University Press, 1999.

Tirion, Wil. *Bright Star Atlas.* Richmond, VA: Willmann Bell, 2001.

———. *Cambridge Star Atlas*, 3rd edition. Cambridge University Press, 2001.

Tirion, Wil, and Roger Sinnott. *Sky Atlas 2000.0*, 2nd edition. Cambridge University Press; Cambridge, MA: Sky Publishing, 1998.

INDEX

Page numbers in **bold type** indicate photographs.